Journal of Approximation Theory and Applied Mathematics

ISSN 2196-1581

Impressum:

Michael Rasguljajew

Kasinostraße 63
64293 Darmstadt
Germany, Hessen

E-Mail: m.rasguljajew@jatam.de

Website: www.jatam.com

Printed and published:
BoD - Books on Demand,
Norderstedt

ISBN 978-3-7392-4662-8

Contents

New Methods of Approximation of Step Functions

S. V. Aliukov

South Ural State University, Chelyabinsk, Russia
e-mail: alysergey@gmail.com

Abstract—New methods of approximation of step functions with an estimation of the error of the approximation are suggested. The suggested methods do not have any of the disadvantages of traditional approximations of step functions by means of Fourier series and can be used in problems of mathematical modeling of a wide range of processes and systems.

Keywords: step functions, mathematical modeling, approximation, convergence, estimation of error, examples of application.

1. INTRODUCTION

Step functions are widely applied in various areas of scientific research. Technical and mathematical disciplines, such as automatic control theory, electrical and radio engineering, information and signal transmission theory, equations of mathematical physics, theory of vibrations, and differential equations are traditional fields of application [1–3].

Systems with step parameters and functions are considered highly nonlinear structures to emphasize the complexity of obtaining solutions for such structures. Despite the simplicity of step functions in segments, the construction of solutions in problems with step functions on the whole domain of definition requires using special mathematical methods, such as the alignment method [4] with the coordination of the solution by segments and switching surfaces. Generally, application of the alignment method requires overcoming substantial mathematical difficulties, and intricate solutions represented by complex expressions are obtained rather often.

In many cases, researchers rely upon approximation methods using Fourier series $f = \sum\limits_{k=1}^{\infty} c_k \varphi_k$, where $\{\varphi_1, \varphi_2, \ldots, \varphi_n, \ldots\}$ is an orthogonal system in functional Hilbert space $L_2[-\pi, \pi]$ of measurable functions with Lebesgue integrable squares, $f \in L_2[-\pi,\pi]$, $c_k = (f \cdot \varphi_k)/\|\varphi_k\|^2$. The trigonometric system of 2π periodic functions $\{1, \sin nx, \cos nx; n \in N\}$ is often taken as an orthogonal system. In this case, the following is fulfilled in the vicinity of discontinuity points $O_\delta(x_0)$ $\sup\limits_{x \in O_\delta(x_0)/\{x_0\}} |f(x) - S_n(x)| \xrightarrow[n \to \infty]{} A \neq 0$, where $S_n(x)$ is the partial sum of the Fourier series. It is how Gibbs' phenomenon shows itself [5]. Thus, in the case of a function

$$f_0(x) = \operatorname{sign}(\sin x) \tag{1}$$

the point $x = \pi/m$, where $m = 2[(n+1)/2]$, and $[A]$ is the integral part of the number A, is the maximum point of the partial sum $S_n(f_0)$ of the trigonometric Fourier series [6] with

$$S_n(f_0, \pi/m) \xrightarrow[n \to \infty]{} \frac{2}{\pi} \int_0^\pi \frac{\sin t}{t} dt \approx 1{,}17898,$$

i.e., the absolute error value $\left| f_0(\pi/m) - \lim\limits_{n\to\infty} S_n(f_0, \pi/m) \right| > 0$. It should be noted that $x = \pi/m \xrightarrow[n\to\infty]{} 0+0$.

The graph of the partial sum $S_{20}(f_0)$ of the trigonometric series on the interval $[-\pi, \pi]$, which illustrates the presence of the Gibbs phenomenon is presented in Fig. 1.

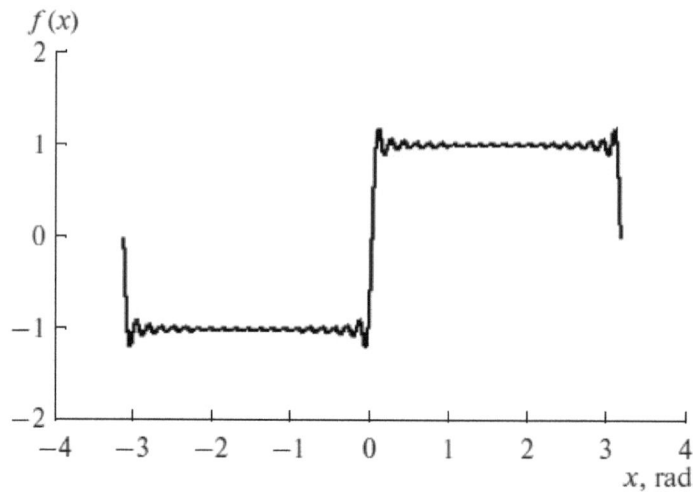

Fig. 1. Presence of the Gibbs phenomenon

What is unpleasant in this case is that the Gibbs effect is generic and is present for any function $f \in L_2[a, b]$, which has limited variation on the interval $[a, b]$, with isolated discontinuity point $x_0 \in (a, b)$. The following condition is fulfilled for such functions [6]

$$\lim_{n\to\infty} S_n(f, x_0 + \pi/m) = f(x_0+0) + \frac{d}{2}\cdot\left(\frac{2}{\pi}\int_0^\pi \frac{\sin t}{t}dt - 1\right), \text{ where } d = f(x_0+0) - f(x_0-0).$$

We show that absolute $\Delta = \Delta(x)$ and relative $\delta = \delta(x)$ errors of approximation in the vicinity of discontinuity points may be as large as we please. In fact,

$$\lim_{n\to\infty} \Delta(x_0 + \pi/m) = \lim_{n\to\infty}\left| S_n(f, x_0 + \pi/m) - f(x_0 + \pi/m)\right| = \left|\lim_{n\to\infty} S_n(x_0 + \pi/m) - \lim_{n\to\infty} f(x_0 + \pi/m)\right| =$$

$$= \left| f(x_0+0) + \frac{d}{2}\cdot\left(\frac{2}{\pi}\int_0^\pi \frac{\sin t}{t}dt - 1\right) - f(x_0+0)\right| = \left|\frac{d}{2}\cdot\left(\frac{2}{\pi}\int_0^\pi \frac{\sin t}{t}dt - 1\right)\right| = \Delta(d).$$

The function $\Delta(d)$ is an infinitely large value, as

$$\forall M > 0 \, \exists d = d^*(M) > 0 \, \forall d : |d| > d^* \Rightarrow \Delta(d^*) = \left|\frac{d^*}{2}\cdot\left(\frac{2}{\pi}\int_0^\pi \frac{\sin t}{t}dt - 1\right)\right| > M.$$ Such expression

as $\left[2M\pi \bigg/ \left(2\int_0^\pi \frac{\sin t}{t}dt - \pi\right)\right] + 1$, where $[A]$ is the integral part of the number A, may be taken as d^*.

The proof is identical for the relative error $\delta(x) = \Delta(x)/|f(x)|$. Moreover, even when $d \in R \ (d \neq 0)$ is fixed for any $M > 0$, the function $f(x) \in L_2[a,b]$ may be selected in such a way that $\delta(x_0+0,d) = \Delta(x_0+0,d)/|f(x_0+0)| > M$. The function with $|f(x_0+0)| < \Delta(x_0+0,d)/M, \ f(x_0+0) \neq 0$ may be taken as an example for this case.

It should be noted that it is not necessary for the Fourier series to converge at each point even on the set of continuous functions $C[-\pi, \pi]$, which is commonly known.

The presence of the Gibbs phenomenon leads to extremely negative consequences of the use of the partial sum of a trigonometric series as an approximating function in fields such as radio engineering and signal transmission.

2. DESCRIPTION OF THE METHOD

In order to eliminate the mentioned disadvantages, new methods of approximation of step functions based on the use of trigonometric expressions represented by recursive functions are suggested in the present paper.

For example, consider the step function (1) in more detail. This function is often used as an example of the application of Fourier series, and, therefore, it is convenient to take this function for comparative analysis of a traditional Fourier series expansion and the suggested method. Expansion of (1) into Fourier series has all the above mentioned disadvantages. In order to eliminate them, it is proposed to approximate the initial step function by a sequence of recursive periodic functions

$$\left\{f_n(x) \,\middle|\, f_n(x) = \sin\!\big((\pi/2)\cdot f_{n-1}(x)\big), f_1(x) = \sin x; n-1 \in N\right\} \subset C^\infty[-\pi, \pi] \qquad (2)$$

Graphs of the initial function (a thickened line) and its five successive approximations for this case are presented in Fig. 2. It can be seen that, even when n values are relatively small in the iterative procedure (2), the graph of the approximating functions approximates the initial function (1) rather well. In addition, approximating functions obtained using the suggested method do not have any of the disadvantages of Fourier series expansion. There is absolutely no sign of the Gibbs phenomenon.

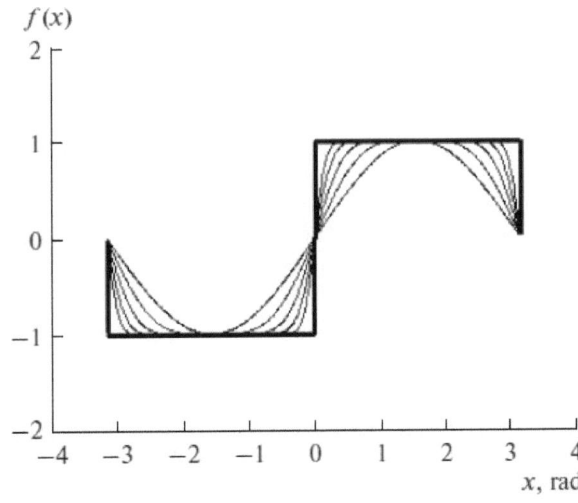

Fig.2 Graphs of the initial function and its successive approximation

Certain peculiarities of the proposed approximating iterative procedure are to be mentioned.

It should be noted that functions $f_n(x)$ and $f_0(x)$ are uneven and periodic ones with a period of 2π. Functions $f_n(x+\pi/2)$ and $f_0(x+\pi/2)$ are even and periodic. Therefore, it is sufficient to consider the sequence of approximating functions (2) on the interval $[0, \pi/2]$.

Let $\{f_n(x)\} \subset L_2[0, \pi/2]$ and $f_0(x) \in L_2[0, \pi/2]$. As $\sup\limits_{n\in N} \sup\limits_{x\in[0,\,\pi/2]} |f_n(x)| = 1 < \infty$ (due to

the boundedness of functions $fn(x)$) and $\sup\limits_{n\in N} \underset{0}{\overset{\pi/2}{Var}}\, f_n = 1 < \infty$ (due to the monotonicity of functions

$f_n(x)$ on the interval $[0, \pi/2]$), then, a subsequence converging at each point of $[0, \pi/2]$ to a certain function f with $\overset{\pi/2}{\underset{0}{Var}} f \leq \overline{\underset{n\to\infty}{\lim}} \overset{\pi/2}{\underset{0}{Var}} f_n$ may be extracted from the sequence $\{f_n(x)\}$ based on Helly's theorem. The possibility of taking the initial function $f_0(x)$ as such function f will be shown below.

Theorem 1. A sequence of functions $f_n(x)$ converges to the initial function $f_0(x)$, with the convergence being point-by-point, though not uniform.

Proof. We have $f_n(x) - f_0(x) = 0$, $\forall n \in N$ at $x = 0$ and $x = \pi/2$. Therefore, $f_n(x) \xrightarrow[n\to\infty]{} f_0(x)$ at these points, as $\forall \varepsilon > 0 \, \exists n^* \in N \, \forall n : n > n^* \Rightarrow |f_n(x) - f_0(x)| < \varepsilon$. We may set $n^* = 1$ as an example.

As $\sin x > (2/\pi) \cdot x, \forall x \in (0, \pi/2)$, then the condition $f_n(x) = \sin((\pi/2) \cdot f_{n-1}(x)) > f_{n-1}(x) > \ldots > f_1(x) > 0$ is fulfilled for any $x \in (0, \pi/2)$. Then, the sequence $f_n(x), \forall x \in (0, \pi/2)$ is positive, ascending, and limited, and, therefore, it has the finite limit, which will be indicated as $\underset{n\to\infty}{\lim} f_n(x) = A \in R$ We obtain $A = \underset{n\to\infty}{\lim} \sin((\pi/2) \cdot f_{n-1}(x)) = \sin((\pi/2) \cdot \underset{n\to\infty}{\lim} f_{n-1}(x)) = \sin((\pi/2) \cdot A)$, based on which we find that $A = 0$ or $A = 1$. As the sequence is of positive terms and ascending, then $A = 1 = f_0(x)$. Then, $f_n(x) \xrightarrow[n\to\infty]{} f_0(x)$ on the considered interval. With the conclusion on convergence of the sequence at $x = 0$ and $x = \pi/2$, which was made above, we conclude that $f_n(x) \xrightarrow[n\to\infty]{} f_0(x), \forall x \in [0, \pi/2]$. This convergence is only a point-by-point one, but not uniform, as the function $f_0(x)$ is not continuous on the interval $[0, \pi/2]$.

Theorem 2. The sequence of approximating functions $f_n(x)$ converges along the norm towards the initial function $f_0(x)$ in Banach $L_1[0, \pi/2]$ and Hilbert spaces of measurable functions $L_2[0, \pi/2]$.

Proof. We introduce the sequence of functions $\{\eta_n(x) \,|\, \eta_n(x) = (2/\pi) \cdot \arctg(n\pi); n \in N\} \subset C^\infty[0, \pi/2]$, which are minorant with respect to the sequence $f_n(x)$. It may be shown that $f_n(x) \geq \eta_n(x), \forall n \in N, \forall x \in [0, \pi/2]$. It should be noted that the measure of the set of discontinuity points of the function $f_0(x)$ is zero.

Then, with functions $f_n(x)$ and $\eta_n(x)$ being non-negative terms and limited on the considered interval, we obtain the following in the space $L_1[0, \pi/2]$:

$$\|f_0(x) - f_n(x)\| = \int_0^{\pi/2}(1 - f_n(x))dx \leq \int_0^{\pi/2}(1 - \eta_n(x))dx = \frac{\pi}{2} - \arctg\frac{\pi n}{2} + \frac{1}{\pi n} \cdot \ln\left(1 + (\pi n)^2/4\right)$$

As $\underset{n\to\infty}{\lim}\left(\frac{\pi}{2} - \arctg\frac{\pi n}{2} + \frac{1}{\pi n} \cdot \ln\left(1 + (\pi n)^2/4\right)\right) = 0$, then $\|f_0(x) - f_n(x)\| \xrightarrow[n\to\infty]{} 0$.

Similarly it may be proved that the sequence $f_n(x)$ converges along the norm towards the function $f_0(x)$ in the space $L_2[0, \pi/2]$.

Thus, the sequence of approximating functions $f_n(x)$ in spaces $L_1[-\pi, \pi]$ and $L_2[-\pi, \pi]$ is fundamental. Whereas, the sequence $f_n(x)$ is not fundamental in the space $C[-\pi, \pi]$.

The number $\pi/2$ was used in the sequence of approximating functions (3) as a constant factor; however, it is possible to take another factor, which may be variable as well. Cosine and other trigonometric functions and their combinations may be used instead of sine in the suggested method of approximation. For example, if we use the sequence of recursive functions

$$\{f_n(x) \mid f_n(x) = \cos(\varphi_n(x)), \varphi_n(x) = (\pi/2) \cdot \sin(\varphi_{n-1}(x)), \varphi_1(x) = x, n-1 \in N\} \subset C^{\infty}[-\pi, \pi],$$

we may approximate short-term impulses. The graph of one function from such sequence is presented in Fig. 3.

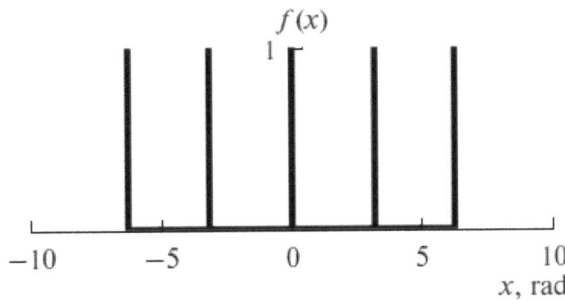

Fig. 3. Graph of the analytical function that approximates short-term pulses

These functions may be used for mathematical models describing the transmission of short-term signals, mechanical systems with shock interactions, etc. It should be noted that, despite the impulse (highly nonlinear) shape of graphs of such functions, they are continuous analytical functions and tolerate the application of analytical methods. The error of the approximation in spaces $L_1[-\pi, \pi]$ and $L_2[-\pi, \pi]$ may be as small as we please in these case.

We return to the sequence of approximating functions (2). The function $f_1(x)$ will be called the initial function (or angle one). We may use another function (not necessarily a periodic one) instead of sine as the initial function. It should be noted that, when iterative procedure (2) is used and given condition $|f_1(x)| < 2$, we obtain $\lim\limits_{n \to \infty} f_n(x) = \text{sign}(f_1(x))$. In addition, we may approximate any step function. In fact, we will take the initial function written as $f_1(x) = \exp(1 - (ax + b)^2) - 1$ for the step function

$$f(x) = \begin{cases} h, \ x \in (x_1, x_2), \\ 0, \ x \notin (x_1, x_2). \end{cases} \tag{3}$$

We obtain $a = 2/(x_1 - x_2); \ b = (x_1 + x_2)/(x_2 - x_1)$ based on the condition $f_1(x_1) = f_1(x_2) = 0$. The sequence

$$\{f_n(x) \mid f_n(x) = (h/2) \cdot (1 + \sin\varphi_n(x)), \varphi_n(x) = (\pi/2) \cdot \sin\varphi_{n-1}, \varphi_1(x) = (\pi/2) \cdot f_1(x), n-1 \in N\}$$

for these values of coefficients a and b converges to the step function $f(x)$. Then, any step function with values $h_i i$ on intervals (x_{1i}, x_{2i}) may be approximated by the sum of identical sequences $\sum\limits_{i=1}^{k} \{f_n(x)\}_i$.

When we considered approximation of the step function $f(x)$ (3), we assumed that its position and height are precisely known. In actual problems parameters are usually set approximately. Let, for example, the initial parameters be set with absolute errors $|\hat{x}_1 - x_1| = \Delta x_1 \in [0, \Delta^* x_1), |\hat{x}_2 - x_2| = \Delta x_2 \in [0, \Delta^* x_2), |\hat{h} - h| = \Delta h \in [0, \Delta^* h)$, where $\Delta^* x_1 = \sup \Delta x_1, \Delta^* x_2 = \sup \Delta x_2, \Delta^* h = \sup \Delta h$, $\hat{x}_1, \hat{x}_2, \hat{h}$ are approximated values of the parameters. We consider step function (3) on the interval $[c, d]$, for which $[x_1 - \Delta^* x_1, x_2 + \Delta^* x_2] \subset [c, d]$. In this case, we obtain the following estimated absolute errors of approximation with respect to the norm in spaces $L_1[c, d], L_2[c, d]$ and $M[c, d]$

respectively, with $M[c, d]$ being the set of functions with metric $\rho(f^{(1)}(x), f^{(2)}(x)) = \sup\limits_{x \in [c,d]} \left| f^{(1)}(x) - f^{(2)}(x) \right|$ limited on the interval $[c, d]$:

$$\| \Delta f \| < \sup\limits_{\Delta x_1} \sup\limits_{\Delta x_2} \sup\limits_{\Delta h} \lim\limits_{n \to \infty} \| f(x) - f_n(x) \|_{L_1[c,d]} = (|h| + \Delta^* h) \cdot (\Delta^* x_1 + \Delta x_2) + (x_2 - x_1) \cdot \Delta^* h;$$

$$\| \Delta f \| < \sup\limits_{\Delta x_1} \sup\limits_{\Delta x_2} \sup\limits_{\Delta h} \lim\limits_{n \to \infty} \| f(x) - f_n(x) \|_{L_2[c,d]} = \sqrt{(|h| + \Delta^* h)^2 \cdot (\Delta^* x_1 + \Delta^* x_2) + (x_2 - x_1) \cdot \Delta^* h^2};$$

$$\| \Delta f \| < \sup\limits_{\Delta x_1} \sup\limits_{\Delta x_2} \sup\limits_{\Delta h} \lim\limits_{n \to \infty} \| f(x) - f_n(x) \|_{M[c,d]} = |h| + \Delta^* h.$$

It can be seen from the obtained estimations that the error of approximation does not accumulate, which is a positive aspect of the suggested method.

As in practice we usually only know the approximate parameter values and measurement errors, it is more convenient to express the upper bound estimates for the absolute error of approximation as follows:

$$\| \Delta f \|_{L_1[c,d]} < (|\hat{h}| + 2\Delta^* h) \cdot (\Delta^* x_1 + \Delta^* x_2) + (\hat{x}_2 - \hat{x}_1 + \Delta^* x_1 + \Delta^* x_2) \cdot \Delta^* h;$$

$$\| \Delta f \|_{L_2[c,d]} < (|\hat{h}| + 2\Delta^* h)^2 \cdot (\Delta^* x_1 + \Delta^* x_2) + (\hat{x}_2 - \hat{x}_1 + \Delta^* x_1 + \Delta^* x_2) \cdot \Delta^* h^2;$$

$$\| \Delta f \|_{M[c,d]} < |\hat{h}| + 2\Delta^* h.$$

We return to (1) and its approximation using (2) in the space of limited functions $M[0, \pi]$.

Let $\Delta = |f_0(x) - f_n(x)| \in [0, 1]$ be the absolute error of approximation. We write down the

sequence $\left\{ r_n \;\middle|\; r_n = \max\limits_{\Delta} \max\limits_{x1, x2 \in [0,\pi]\,:\, f_n(x1) = f_n(x2)} |x2 - x1| \right\}$ of maximum metrics. Based on the

equation $f_n(x) = 1 - \Delta$, we find that this sequence may be represented as $\left\{ r_n \;\middle|\; r_n = \pi - 2\arcsin \lambda_n, \lambda_n = (2/\pi) \cdot \arcsin \lambda_{n-1}, \lambda_1 = 1 - \Delta, n-1 \in N \right\}$. Similarly to the proof of

the theorem 1, it may be proved that $r_n(\Delta) \xrightarrow[n \to \infty]{} r^*(\Delta) = \begin{cases} \pi, \Delta \in (0, 1], \\ 0, \Delta = 0, \end{cases}$

with convergence on the interval $[0, 1]$ being point-by-point without being uniform. It is important to mention that the sequence $\{r_n\}$ converges to the step function as well.

The graphs of several of the initial functions from the sequence $r_n(\Delta)$ are presented in Fig. 4. It can be seen from Fig. 4 that the length of the interval on which the error of approximation does not exceed Δ rises sharply when n is increased in the area of rather small values of the error Δ. This proves the rapid convergence of the suggested method and is its positive peculiarity.

In order to quantitatively estimate the change in the length of this interval, we deduce the approximate dependence for the function $\Delta r(n, \Delta) = r_n - r_{n-1}$. To do this, we use the relationship $r_n - r_{n-1} = 2(x_{n-1} - x_n)$, where $x_n = \arcsin((2/\pi) \cdot x_{n-1})$, $x_1 = \arcsin(1 - \Delta)$. Then we get $r_n - r_{n-1} = 2(x_{n-1} - \arcsin((2/\pi) \cdot x_{n-1}))$. When we expand $\arcsin((2/\pi) \cdot x_{n-1})$ into the Maclaurin series and consider that x_{n-1} values are rather small, we obtain approximately the following formula: $r_n - r_{n-1} \approx (2/\pi) \cdot (\pi - 2) \cdot x_{n-1}$. Then $r_n - r_{n-1} \approx (2/\pi)^{n-1} \cdot (\pi - 2) \cdot \arcsin(1 - \Delta)$.

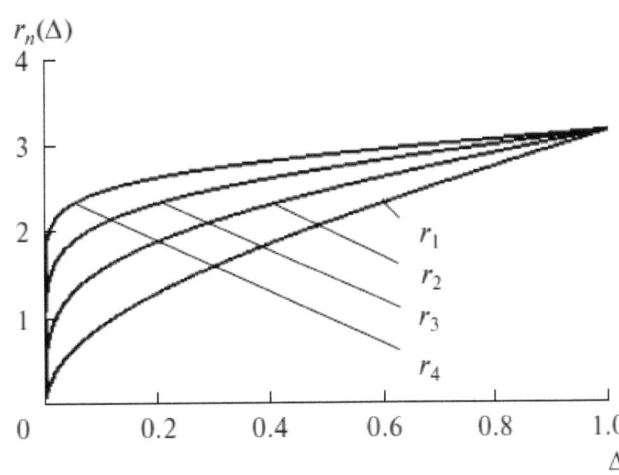

Fig. 4. Lengths of intervals with the error of approximation not exceeding Δ

We indicate several properties of the suggested approximation (2).

Property 1. The maximum difference in lengths of intervals $r_n - r_{n-1}$ does not depend on n and is obtained using the relationship

$$\max_{\Delta \in [0,1]} (r_n - r_{n-1}) = \sqrt{\pi^2 - 4} - 2\arcsin\sqrt{1 - 4/\pi^2}, \; n\text{-}1 \in N.$$

Proof. Based on the previously obtained relationship $r_n - r_{n-1} = 2(x_{n-1} - \arcsin((2/\pi) \cdot x_{n-1})), \; n\text{-}1 \in N,$ we obtain the derivative

$$\frac{d(r_n - r_{n-1})}{d\Delta} = -2^{n-1} \cdot \left(\sqrt{\pi^2 - 4x_{n-1}^2} - 2\right) \Big/ \sqrt{\prod_{i=1}^{n-1} (\pi^2 - 4x_i^2)(1 - (1-\Delta)^2)}.$$

The points $x_{n-1} = x_{n-2} = \ldots = x_1 = \pi/2$ are minimum points, at which $r_n - r_{n-1} = 0$. We also obtain $r_n - r_{n-1} = 0$ in the case of $\Delta = 1$. The points $x_{n-1} = \sqrt{(\pi^2/4) - 1}$ are maximum points and do not depend on n. Then, we obtain

$$\max_{\Delta \in [0,1]} (r_n - r_{n-1}) = \sqrt{\pi^2 - 4} - 2\arcsin\sqrt{1 - 4/\pi^2}.$$ It is indicated that

$$\max_{\Delta \in [0,1]} (r_n - r_{n-1}) \approx 0{,}661 \text{ for use as a reference.}$$

Property 2. The maximum difference in the values of functions $f_n(x) - f_{n-1}(x)$ does not depend on n and is obtained using the relationship

$$\max_{x \in [0, \pi]} (f_n(x) - f_{n-1}(x)) = (\sqrt{\pi^2 - 4} - 2\arccos(2/\pi))/\pi, \; n\text{-}1 \in N.$$

The proof is similar to the one for property 1. It is also indicated that $\max_{x \in [0, \pi]} (f_n(x) - f_{n-1}(x)) \approx 0{,}211$ for use as a reference.

Property 2 shows that the sequence of approximating functions $f_n(x)$ (2) is not Cauchy convergent; i.e., it is not fundamental, as $\exists \varepsilon > 0 \; \forall n^* \in N \; \exists n, m > n^*$, which is $\max_{x \in [0,\pi]} |f_n(x) - f_m(x)| > \varepsilon$. The number 0.1 may be taken as ε, for example, given that

$$m = n^* + 1, n = n^* + 2.$$

The obtained relationships may be used to estimate errors of approximation in the solution of the applied problems.

3. EXAMPLES OF APPLICATION

The sequence of sine mechanisms may act as a mechanical analog of the suggested approximations [7]. For example, the sequence of sine mechanisms in a position that corresponds to approximation (3) is presented in Fig. 5. Here, (*1*) represents the drive shaft; (*2*) represents the crank; (*3*) is the slide block; (*4*) is the link; (*5*) represents the rack; and (*6*) is the gear wheel. Then, the sequence of elements is repeated (it may be repeated several times). Element *7* corresponds to the output element, which may be connected to the drive shaft through the pinion.

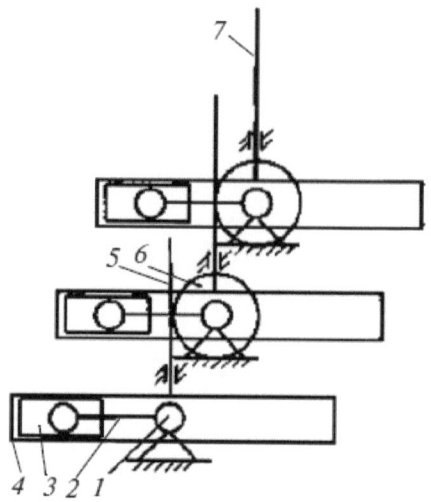

Fig. 5. Mechanical analog of the approximation of a step function

The mechanism presented in Fig. 5 transforms the uniform rotational motion of drive shaft *1* into the intermittent reciprocal or vibrational motion of the output element (with any degree of accuracy). In addition, the different relative positions of the cranks and the different relationships of the sizes of the crank and gear wheels make it possible to simulate various laws of motion of the output element that correspond to the considered approximations of the step functions (discontinuous motions, impulsive motions, etc.). Such a mechanism may be applied, for example, as transport mechanism in tape drive systems to provide a higher quality of the execution of the process. It is also possible to apply such a mechanism in pulse variators to achieve a more uniform motion of the output drive shaft, as the vibrational processes occur in accordance with the curves composed of the segments being approximated to a constant, rather than with a sine wave.

The suggested method of approximation makes it possible to obtain functions which may be applied, for example, in the design and manufacturing of gear wheels and spline joints. The result of the construction of gear profile in MathCAD software is presented in Fig. 6. The function $r(\varphi) = A + a \cdot \sin((\pi/2) \cdot \sin(n\varphi))$, where A is the pitch circle radius, a is the tooth point height, n is the number of teeth, and r and φ are polar radius and polar angle, respectively, was taken as a basis for the construction of the profile.

When we change the number of nested trigonometric functions used for approximation and vary their parameters, we may get gear profiles and spline joints with enhanced reliability in comparison with evolvent ones, which, in contrast with rectangular splines, for example, have no significant stress concentrators.

We consider one more example of the application of the developed methods of approximation. The optical installation for sound reconstruction from worn and damaged records, which makes it possible to obtain electronic profiles of sound carrying media via a noncontact method, was developed in the United States (Figs. 7a, 7b) [8].

Several tens (sometimes even hundreds) of thousands of electronic radial profiles of each sound carrying medium being reconstructed are acquired using the optical installation. Approximation, mostly the one using linear splines (Fig. 7c), is used for the reconstruction of electronic profiles. The accuracy of the approximation leaves much to be desired in this case, whereas the electronic analog of the initial profile may be reconstructed with a rather high degree of accuracy using the methods suggested in the present paper (Fig. 7e).

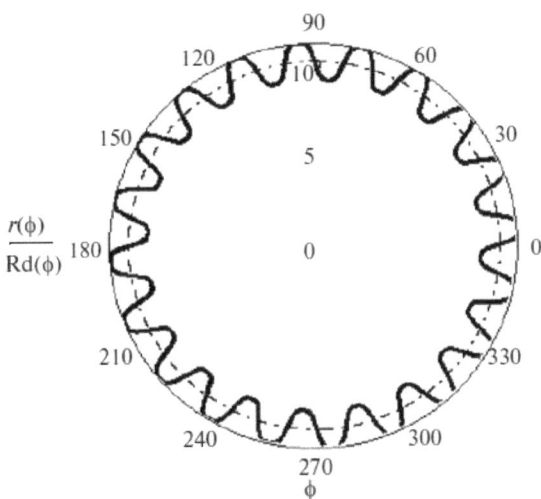

Fig.6. Construction of a gear profile using the approximation of step functions

The black dots in Fig. 7d indicate the measurements of the radial profile of sound tracks using the optical installation. An enlarged image of Fig. 7d is presented in Fig. 7e, where the white line corresponds to an approximation of the profile using the suggested technique.

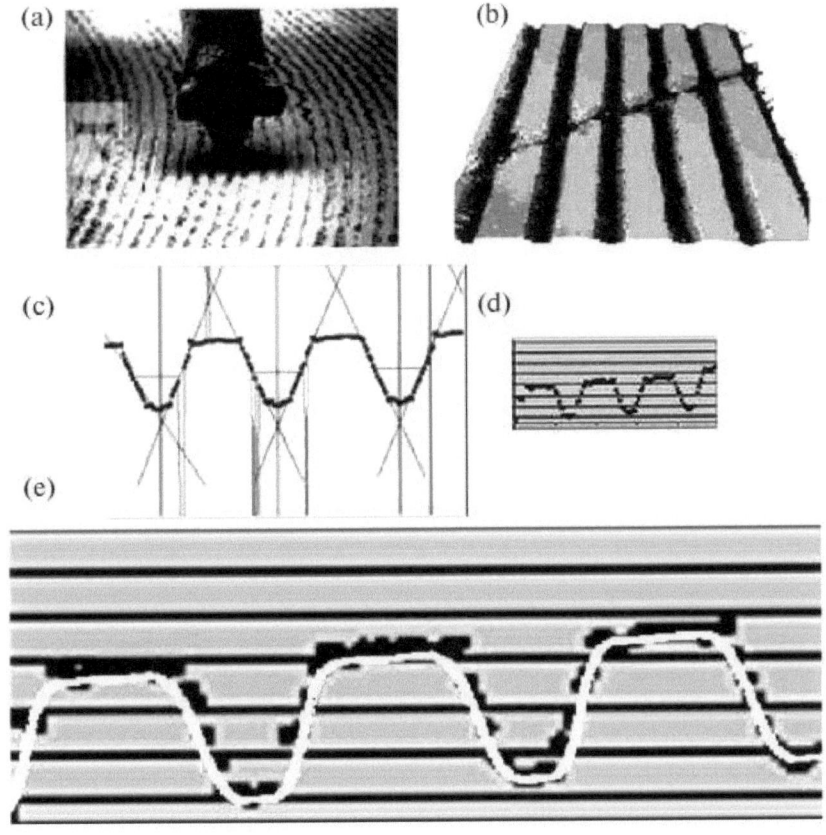

Fig. 7. Profiles of sound tracks and their application

The described methods of approximation make it possible to automate the process of the reconstruction of electronic sound tracks, which is very important for the execution of the process with a high performance level, taking into consideration the significant number (tens and hundreds of thousands) of electronic radial profiles.

The suggested methods of approximation may also be used in the mathematical modeling of biomedical processes. For example, a fragment of a cardiogram is presented in Fig. 8. Approximation is carried out for one of the graphs of the cardiogram using the proposed procedure. The graphic results of the approximation by means of one of the developed functions are presented in an enlarged form in the middle part of the figure (Fig. 8a), where the approximating function is superimposed on the graph of the cardiogram. In order to better understand the graph of the approximating function, this graph in Fig. 8b is presented in a position shifted with respect to the cardiogram. It can be seen that the approximation is quite accurate. Similar approximations may be carried out for other graphs of the cardiogram as well.

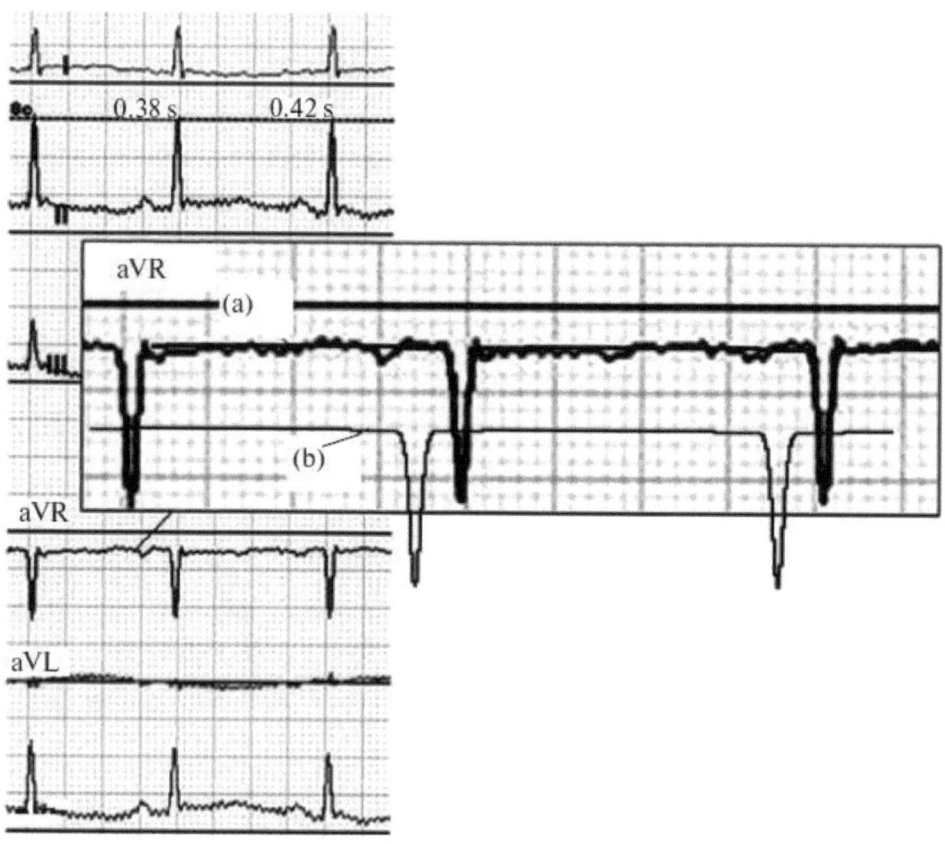

Fig. 8. Approximation of a fragment of the cardiogram

The possibility of using the proposed methods for approximation of nonperiodic step functions should also be noted. The period of approximating functions in this case should be rather large covering the area of possible values of the argument of the function being approximated of the actual process being investigated. Such an approximation may be used in the modeling of, for example, technical systems with dry friction parameters and inertial transformer dynamics with an alternating -sign moment of resistance.

4. NUMERICAL TESTING

Numerical testing of the proposed approximating procedure will be carried out using the example of investigation of dynamics of an inertial impulse stepless gear. It is known [9] that weak elements (free wheel mechanisms) may be excluded from the construction of the inertial impulse steplsess gear based on planetary gear with unbalanced satellites under condition that the moment of resistance affecting the drive shaft has an alternating sign. The gear's dynamics may be described by the highly nonlinear second-order differential equation

$$A_1 \ddot{\beta} + A_2(\omega - \dot{\beta})^2 - A_3 \omega^2 = -M_C,$$

where
$A_1 = B_1 + b_1 \cdot \cos\psi,\ A_2 = a_2 \cdot \sin\psi,\ A_3 = a_3 \cdot \sin\psi,\ \psi = q(\omega \cdot t - \beta),$

B_1, b_1, a_2, a_3, q are constant coefficients, including the gear parameters,

$M_C = M_1 \cdot \text{sign}(\dot{\beta}) + M_0$ is the moment of resistance affecting the drive shaft $(M_0, M_1 \equiv const)$,

$\omega \equiv const$ is the angular velocity of the driveshaft,

β is the rotation angle of the driven shaft, and $\bullet \equiv \dfrac{d}{dt}$ is the operator of derivation with respect to time t.

The sign function $\text{sign}(\dot{\beta})$ is highly nonlinear, which complicates carrying out analytical investigations of the dynamics of the inertial impulse gear. In addition, this function is not periodic. We approximate the sign function using the suggested methods (2) by, for example, an analytical function written as $sign(\dot{\beta}) \approx f_4(\dot{\beta}/10)$. It should be noted that we take relatively small $n = 4$ for the approximation, leaving substantial opportunities for a reduction in the approximation error.

For the sake of comparison, we carry out a numerical solution of the differential motion equation with the sign and the approximating functions for particular examples of gears according to the Runge–Kutta method. Phase trajectories on phase plane $(\beta, \dot{\beta})$ with access to a periodic solution are presented in Fig. 9. Here, the solid line indicates the solution obtained with the gear with a discontinuous sign function used in the mathematical model, while the dotted line represents the solution obtained using an analytical approximation. The thickened line in Fig. 9 corresponds to the periodic solution.

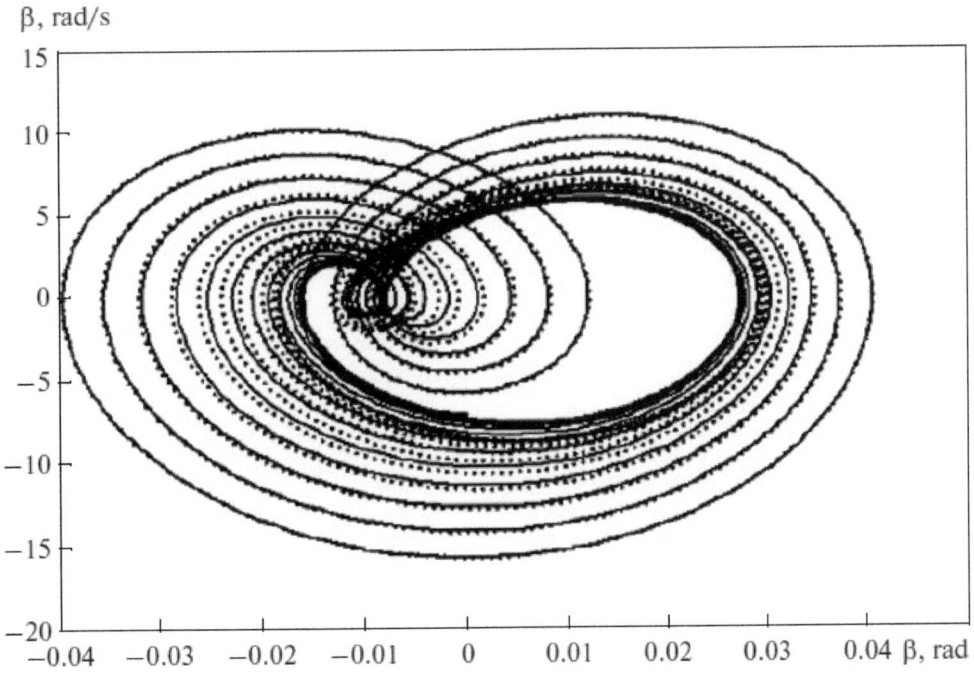

Fig. 9. Phase trajectories in the case of use of the sign function and its approximation

It can be seen from the figure that the error of the results is not large, which shows good convergence of the suggested approximating procedures. Furthermore, the approximation error may be reduced to as small value as desired through an increase in the number of nested functions.

The considered examples are taken from various areas and are not single ones for the application of the suggested methods of approximation. Therefore, sufficient universality of these methods may be stated.

The described methods of approximation do not have any of the disadvantages of expansions of functions into Fourier series and may find wide use in the solution of applied problems. It should also be noted that the proposed approximating functions are continuous and analytical ones. They reflect actual processes to a larger extent than step functions, as even jump processes occur in reality within short, but not zero, time intervals.

REFERENCES

1. A. V. Nikitin and V. F. Shishlakov, *Parametric Synthesis of Nonlinear Automatic Control Systems* (SPbGUAP, St. Petersburg, 2003) [in Russian].

2. D. Meltzer, "On the Expressibility of Piecewise Linear Continuous Functions as the Difference of Two Piece-wise Linear Convex Functions," Math. Program., Study **29**, 118–134 (1986).

3. S. I. Baskakov, *Radio Engineering Chains and Signals: Textbook for Higher-Education Institutions.*, 3rd ed. (Vysshaya shkola, Moscow, 2000) [in Russian].

4. E. P. Popov, *Theory of Nonlinear Automatic Regulation and Control Systems: Textbook*, 2nd ed. (Nauka. Gl. red. fiz.-mat. lit, Moscow, 1988) [in Russian].

5. G. Helmberg, "The Gibbs Phenomenon for Fourier Interpolation," J. Approx. Theory **78**, 41–63 (1994).

6. V. V. Zhuk and G. I. Natanson, *Trigonometric Fourier Series and Approximation Theory Elements* (Leningr. Gos. Univ., Leningrad, 1983) [in Russian].

7. N. V. Gulia, V. G. Klokov, and S. A. Yurkov, *Details of Machines* (Izdatel'skii tsentr "Akademiya", Moscow, 2004) [in Russian].

8. V. Fadeyev and C. Haber, "Reconstruction of Recorded Sound from an Edison Cylinder Using Three_Dimensional Non_Contact Optical Surface Metrology," J. Audio Eng. Soc (2005).

9. S. V. Alyukov, *Dynamics of Inertial Transformer of Rotational Moment without Free Wheel Mechanisms* (VPI, Vladimir, 1983) [in Russian].

Resolution of Nonlinear Partial Differential Equations by Elzaki Transform Decomposition Method

Djelloul ZIANE and Mountassir HAMDI CHERIF
Laboratory of mathematics and its applications (LAMAP),
University of Oran1, P.O. Box 1524, Oran, 31000, Algeria.
Email: djeloulz@yahoo.com; mountassir27@yahoo.fr

November 23, 2015

Abstract

The aim of this work is to extend the application of Elzaki transform decomposition method suggested by M. Khalid et al. to resolve nonlinear partial differential equations. We apply the proposed method to obtain approximate analytical solutions of the proposed problems. Comparison between the numerical and the exact solutions revealed that (ETDM) is an alternative analytical method for solving nonlinear partial differential equations.

Keywords: Adomian decomposition method , Elzaki transform method, nonlinear partial differential equations.
2010MSC: 35A08; 44A05; 35G25.

1   Introduction

There is no secret to the researcher in the field of nonlinear partial differential equations, that the solution of this class of equations is not easy. So we find that many researchers have done and are still doing great efforts to find methods to solve this type of equations. These efforts resulted in the consolidation of this research field in many methods, among them we find the Adomian decomposition method (ADM) ([1]-[5]), variational iteration method (VIM) ([6]-[10]) and homotopy perturbation method (HPM) ([11]-[15]), which have become known in a large number of researchers in this area. A new option emerged recently, includes the composition of Laplace transform, Sumudu transform or Elzaki transform with these methods. Among wich are the Adomian decomposition method coupled with Laplace transform method ([16], [17]), Adomian decomposition Sumudu transform method [18], variational iteration method coupled with Laplace transform method ([19], [20]), variational iteration Sumudu transform method [21], homotopy perturbation transform method [22], homotopy perturbation Sumudu transform method ([23]-[25]), homotopy perturbation Elzaki transform method [27], Elzaki transform decomposition algorithm [29].

The basic motivation of the present study is to extend the application of the Elzaki transform decomposition algorithm suggested in [29] to solve nonlinear partial differential equations. The advantage of this method is its capability of combining two powerful methods for obtaining exact solutions for nonlinear equations. Several examples are given to re-confirm the effectiveness of this method.

2   Basic definitions of ELzaki Transform

A new integral transform called ELzaki transform ([26]-[28]) defined for functions of exponential order, is proclaimed. We consider functions in the set A defined by,

$$A = \left\{ f(t)/M, \ k_1, k_2 > 0, \ |f(t)| < Me^{\frac{|t|}{k_j}}, \ if \ t \in (-1)^j \times [0, \ \infty) \right\}.$$

Definition 1 *If $f(t)$ is function defined for all $t \geqslant 0$, its Elzaki transform is the integral of $f(t)$ times $e^{-\frac{t}{s}}$ from $t = 0$ to ∞. It is a function of s and is defined by $E[f]$*

$$E\left[f(t)\right] = T(s) = s \int\limits_{0}^{\infty} f(t)e^{-\frac{t}{s}}dt.$$

Theorem 2 *ELzaki transform amplifies the coefficients of the power series function,*

$$f(t) = \sum_{n=0}^{\infty} a_n t^n. \tag{1}$$

On the new integral transform "ELzaki Transform"
Is

$$E[f(t)] = T(v) = \sum_{n=0}^{\infty} n! a_n v^{n+2}.\tag{2}$$

Theorem 3 *Let $f(t)$ be in A and Let $T_n(v)$ denote ELzaki transform of nth derivative, $f^{(n)}(t)$ of $f(t)$, then for $n \geq 1$,*

$$T_n(v) = \frac{T(v)}{v^n} - \sum_{k=0}^{n-1} v^{2-n+k} f^{(k)}(0).\tag{3}$$

To obtain Elzaki transform of partial derivative we use integration by parts, and then we have

$$\begin{aligned}E\left(\frac{\partial f(x,t)}{\partial t}\right) &= \tfrac{1}{v} T(x,v) - v f(x,0),\\ E\left(\frac{\partial^2 f(x,t)}{\partial t^2}\right) &= \tfrac{1}{v^2} T(x,v) - f(x,0) - v\frac{\partial f(x,0)}{\partial t},\end{aligned}\tag{4}$$

Properties of Elzaki transform can be found in Refs. ([26], [27]), we mention only the following

1. $E(1) = v^2$;
2. $E(t) = v^3$;
3. $E(t^n) = n! v^{n+2}$;
4. $E^{-1}(v^{n+2}) = \frac{t^n}{n!}$.

3   Elzaki Transform Decomposition Method for PDEs

In this section, we extend the proposed method [29] to solve ordinary differential equations, a new modified method for solving partial differential equations. To illustrate the basic idea of this method, we consider a general non-linear nonhomogeneous partial differential equation

$$\frac{\partial^m u(x,t)}{\partial t^m} + Ru(x,t) + Nu(x,t) = g(x,t),\tag{5}$$

where $m = 1, 2, 3$, with the initial conditions

$$\left.\frac{\partial^{m-1} u(x,t)}{\partial t^{m-1}}\right|_{t=0} = f_{m-1}(x), \qquad m = 1, 2, 3,\tag{6}$$

where $\frac{\partial^m u(x,t)}{\partial t^m}$ is the partial derivative of the function $u(x,t)$ of order m ($m = 1,2,3$), R is the linear differential operator, N represents the general nonlinear differential operator, and $g(x,t)$ is the source term.

Applying the Elzaki Transform (denoted in this paper by E) on both sides of Eq. (5), we get

$$E\left[\frac{\partial^m u(x,t)}{\partial t^m}\right] + E\left[Ru(x,t)\right] + E\left[Nu(x,t)\right] = E\left[g(x,t)\right]. \qquad (7)$$

Using the properties of Elzaki Transform, we obtain

$$v^{-m}E\left[u(x,t)\right] = \sum_{k=0}^{m-1} v^{2-m+k}\frac{\partial^k u(x,0)}{\partial t^k} + E\left[g(x,t)\right] - E\left[Ru(x,t) + Nu(x,t)\right],$$
$$\qquad (8)$$

where $m = 1,2,3$.

And thus, we have

$$E\left[u(x,t)\right] = \sum_{k=0}^{m-1} v^{2+k}\frac{\partial^k u(x,0)}{\partial t^k} + v^m E\left[g(x,t)\right] - v^m E\left[Ru(x,t) + Nu(x,t)\right].$$
$$\qquad (9)$$

Operating the inverse transform on both sides of Eq. (9), we get

$$u(x,t) = G(x,t) - E^{-1}\left(v^m E\left[Ru(x,t) + Nu(x,t)\right]\right), \qquad (10)$$

where $G(x,t)$, represents the term arising from the source term and the prescribed initial conditions.

The second step in Elzaki Transform Decomposition Method, is that we represent the solution as an infinite series given below

$$u(x,t) = \sum_{n=0}^{\infty} u_n(x,t), \qquad (11)$$

and the nonlinear term can be decomposed as

$$Nu(x,t) = \sum_{n=0}^{\infty} A_n, \qquad (12)$$

where A_n are Adomian polynomials [30] of $u_0, u_1, u_2, ..., u_n$ and it can be calculated by the formula given below

$$A_n = \frac{1}{n!}\frac{\partial^n}{\partial\lambda^n}\left[N\left(\sum_{i=0}^{\infty}\lambda^i u_i\right)\right]_{\lambda=0}, \quad n = 0, 1, 2, \cdots. \tag{13}$$

Substituting (11) and (12) in (10), we have

$$\sum_{n=0}^{\infty} u_n = G(x,t) - E^{-1}\left[u^m E\left[R\sum_{n=0}^{\infty} u_n + \sum_{n=0}^{\infty} A_n\right]\right]. \tag{14}$$

On comparing both sides of the Eq. (14), we get

$$\begin{aligned}
u_0(x,t) &= G(x,t), \\
u_1(x,t) &= -E^{-1}\left[u^m E\left[Ru_0(x,t) + A_0\right]\right], \\
u_2(x,t) &= -E^{-1}\left[u^m E\left[Ru_1(x,t) + A_1\right]\right], \\
u_3(x,t) &= -E^{-1}\left[u^m E\left[Ru_2(x,t) + A_2\right]\right], \\
&\quad\vdots
\end{aligned} \tag{15}$$

In general, the recursive relation is given as

$$u_{n+1}(x,t) = -E^{-1}\left[u^m E\left[Ru_n(x,t) + A_n\right]\right], \tag{16}$$

where $m = 1, 2, 3$, and $n \geqslant 0$.

Finally, we approximate the analytical solution $u(x,t)$ by truncated series

$$u(x,t) = \lim_{N\to\infty}\sum_{n=0}^{N} u_n(x,t). \tag{17}$$

The above series solutions generally converge very rapidly [31].

4   Application of the ETDM

In this section, we apply Elzaki transform decomposition method for PDEs to solve nonlinear partial differential equations of the first, second and third order.

Example 4.1
First, we consider the following nonlinear partial differential equation

$$u_t + uu_x - u_{xx} = 0, \tag{18}$$

with initial condition

$$u(x, 0) = x. \tag{19}$$

Applying the Elzaki transform on both sides of Eq. (18). Thus, we get

$$E\left[\, u_t\right] + E\left[uu_x\right] - E\left[u_{xx}\right] = 0. \tag{20}$$

We use the properties of Elzaki transform, we have

$$E\left[\, u(x, t)\right] = xv^2 - vE\left[uu_x - u_{xx}\right]. \tag{21}$$

Taking the inverse Elzaki transform on both sides of Eq. (21), we obtain

$$u(x, t) = x - E^{-1}[vE\left[uu_x - u_{xx}\right]]. \tag{22}$$

By applying the aforesaid decomposition method, we have

$$\sum_{n=0}^{\infty} u_n(x, t) = x - E^{-1}\left[vE\left[\sum_{n=0}^{\infty} A_n(u) - \sum_{n=0}^{\infty}(u_n)_{xx}\right]\right] \tag{23}$$

On comparing both sides of Eq. (23), we get

$$\begin{aligned}
u_0(x, t) &= x, \\
u_1(x, t) &= -E^{-1}\left[vE\left[A_0(u) - u_{0xx}(x, t)\right]\right], \\
u_2(x, t) &= -E^{-1}\left[vE\left[A_1(u) - u_{1xx}(x, t)\right]\right], \\
u_3(x, t) &= -E^{-1}\left[vE\left[A_2(u) - u_{2xx}(x, t)\right]\right], \\
&\vdots
\end{aligned} \tag{24}$$

The first few components of $A_n(u)$ polynomials [30], for example, are given by

$$\begin{aligned}
A_0(u) &= u_0 u_{0x}, \\
A_1(u) &= u_0 u_{1x} + u_1 u_{0x}, \\
A_2(u) &= u_0 u_{2x} + u_2 u_{0x} + u_1 u_{1x}, \\
&\vdots
\end{aligned} \tag{25}$$

Using He's polynomials (25) and the iteration formulas (24) we obtain

$$\begin{aligned}
u_0(x, t) &= x, \\
u_1(x, t) &= -xt, \\
u_2(x, t) &= xt^2, \\
u_3(x, t) &= -xt^3, \\
&\vdots
\end{aligned} \tag{26}$$

The first four terms of the decomposition series solution for Eq. (18) is given by

$$u(x,t) = x - xt + xt^2 - xt^3 + \cdots \qquad (27)$$

That gives

$$u(x,t) = \frac{x}{1+t}, \quad |t| < 1, \qquad (28)$$

which is an exact solution to the KdV equation as presented in [32].

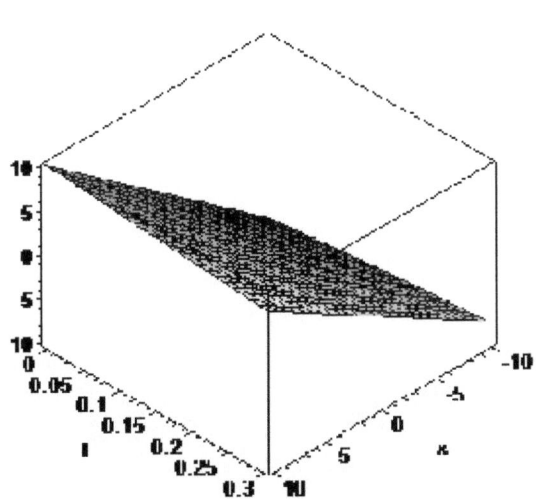

Exact solution (28) of Eq. (18-19).

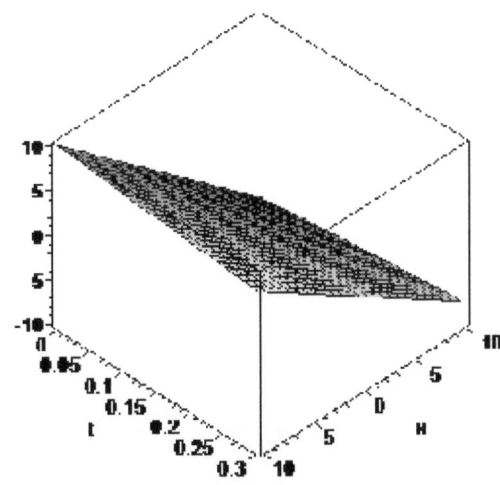

Approximation solution of Eq. (18-19) by ETDM.

Example 4.2

Next, we consider a nonlinear partial differential equation of second order

$$u_{tt} - 2\frac{x^2}{t}uu_x = 0, \ t > 0, \qquad (29)$$

with the initial conditions

$$u(x,0) = 0, \quad u_t(x,0) = x. \qquad (30)$$

The exact solution of this equation is given by

$$u(x, y) = \tan(xt). \tag{31}$$

Applying the Elzaki transform and its inverse on both sides of Eq. (29) gives the following

$$u(x, t) = xt + 2E^{-1}[v^2 E \left[\frac{x^2}{t} u u_x\right]]. \tag{32}$$

By applying the aforesaid decomposition method, we have

$$\sum_{n=0}^{\infty} u_n(x, y) = xt + 2E^{-1}\left[v^2 E \left[\frac{x^2}{t} \sum_{n=0}^{\infty} A_n(u)\right]\right]. \tag{33}$$

On comparing both sides of Eq. (33), we get

$$\begin{aligned} u_0(x, t) &= xt, \\ u_1(x, t) &= 2E^{-1}\left[v^2 E \left[\frac{x^2}{t} A_0(u)\right]\right], \\ u_2(x, t) &= 2E^{-1}\left[v^2 E \left[\frac{x^2}{t} A_1(u)\right]\right], \\ u_3(x, t) &= 2E^{-1}\left[v^2 E \left[\frac{x^2}{t} A_2(u)\right]\right], \\ &\vdots \end{aligned} \tag{34}$$

Using the iteration formulas (34) and He's polynomials (25), we obtain

$$\begin{aligned} u_0(x, t) &= xt, \\ u_1(x, t) &= \tfrac{1}{3}x^3 t^3, \\ u_2(x, t) &= \tfrac{2}{15}x^5 t^5, \\ u_3(x, t) &= \tfrac{17}{315}x^7 t^7. \end{aligned} \tag{35}$$

The approximate solution in a serie form is given by

$$u(x, t) = xt + \frac{1}{3}(xt)^3 + \frac{2}{15}(xt)^5 + \frac{17}{315}(xt)^7 + \circ(xt)^8. \tag{36}$$

And so, we get the exact solution of Eq. (29) that is given in the form

$$u(x, t) = \tan(xt), \tag{37}$$

which is an exact solution to the nonlinear partial differential equation (29) of second order.

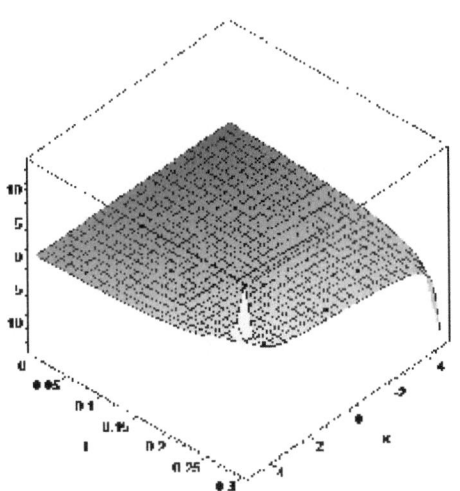

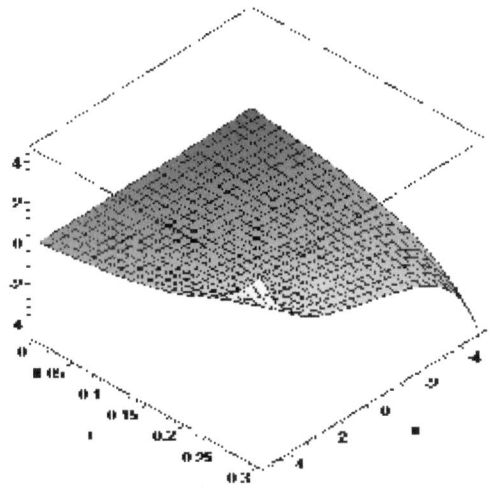

Exact solution (31) of Eq. (29-30).

Approximation solution of Eq. (29-30) by ETDM.

Example 4.3

Finaly, we consider a nonlinear partial differential equation of third order

$$u_{ttt} - 3uu_{xx} = 0, \ t > 0, \tag{38}$$

with the initial conditions

$$u(x,0) = \frac{1}{x}, \ u_t(x,0) = \frac{1}{x^2}, \ u_{tt}(x,0) = \frac{2}{x^3}. \tag{39}$$

The exact solution of the equation (38), is given by

$$u(x,y) = \frac{1}{x-t}, \ \left|\frac{t}{x}\right| < 1, \ x \neq 0. \tag{40}$$

Applying the Elzaki transform and its inverse on both sides of Eq. (38) gives the following

$$u(x,t) = \frac{1}{x} + \frac{t}{x^2} + \frac{t^2}{x^3} + 3E^{-1}[v^3 E[uu_{xx}]]. \tag{41}$$

By applying the aforesaid decomposition method, we have

$$\sum_{n=0}^{\infty} u_n = \frac{1}{x} + \frac{t}{x^2} + \frac{t^2}{x^3} + 3E^{-1}\left[u^3 E\left[\sum_{n=0}^{\infty} A_n(u)\right]\right]. \qquad (42)$$

On comparing both sides of Eq. (42), we get

$$
\begin{aligned}
u_0(x,t) &= \tfrac{1}{x} + \tfrac{t}{x^2} + \tfrac{t^2}{x^3}, \\
u_1(x,t) &= 3E^{-1}\left[v^3 E\left[A_0(u)\right]\right], \\
u_2(x,t) &= 3E^{-1}\left[v^3 E\left[A_1(u)\right]\right], \\
u_3(x,t) &= 3E^{-1}\left[v^3 E\left[A_2(u)\right]\right], \\
&\vdots
\end{aligned} \qquad (43)
$$

The first few components of $A_n(u)$ polynomials [30], for example, are given by

$$
\begin{aligned}
A_0(u) &= u_0 u_{0xx}, \\
A_1(u) &= u_0 u_{1xx} + u_1 u_{0xx}, \\
A_2(u) &= u_0 u_{2xx} + u_1 u_{1xx} + u_2 u_{0xx}, \\
&\vdots
\end{aligned} \qquad (44)
$$

Using the iteration formulas (43) and He's polynomials (44), we obtain

$$
\begin{aligned}
u_0(x,t) &= \tfrac{1}{x}\left[1 + \tfrac{t}{x} + \tfrac{t^2}{x^2}\right], \\
u_1(x,t) &= \tfrac{1}{x}\left[\tfrac{t^3}{x^3} + \tfrac{t^4}{x^4} + \tfrac{t^5}{x^5} + \tfrac{9}{20}\tfrac{t^6}{x^6} + \tfrac{6}{35}\tfrac{t^7}{x^7}\right], \\
u_2(x,t) &= \tfrac{1}{x}\left[\tfrac{11}{20}\tfrac{t^6}{x^6} + \tfrac{29}{35}\tfrac{t^7}{x^7} + \tfrac{t^8}{x^8} + \tfrac{15}{28}\tfrac{t^9}{x^9} + \tfrac{3}{20}\tfrac{t^{10}}{x^{10}}\right], \\
&\vdots
\end{aligned} \qquad (45)
$$

The first terms of the approximate solution of Eq. (38), is given by

$$
\begin{aligned}
U(x,t) = \frac{1}{x}\Bigg[1 &+ \left(\frac{t}{x}\right)^1 + \left(\frac{t}{x}\right)^2 + \left(\frac{t}{x}\right)^3 + \left(\frac{t}{x}\right)^4 + \left(\frac{t}{x}\right)^5 + \left(\frac{t}{x}\right)^6 \\
&+ \left(\frac{t}{x}\right)^7 + \left(\frac{t}{x}\right)^8 + \frac{15}{28}\left(\frac{t}{x}\right)^9 + \frac{3}{20}\left(\frac{t}{x}\right)^{10} + \cdots\Bigg].
\end{aligned} \qquad (46)
$$

That gives

$$U(x,t) = \frac{1}{x-t}, \quad \left|\frac{t}{x}\right| < 1, \; x \neq 0, \qquad (47)$$

which is an exact solution to the nonlinear partial differential equation (38) of third order.

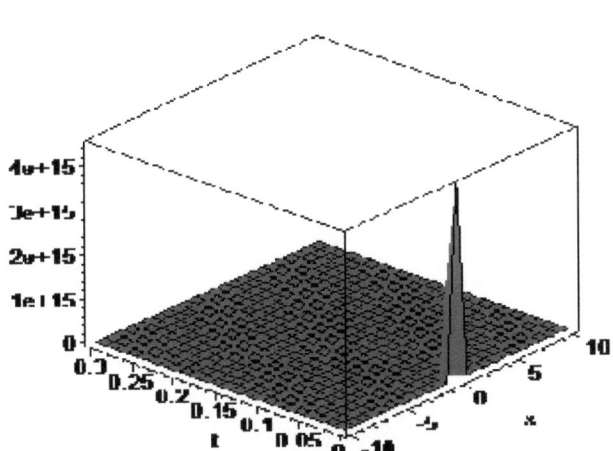

Exact solution (40) of Eq. (38-39).

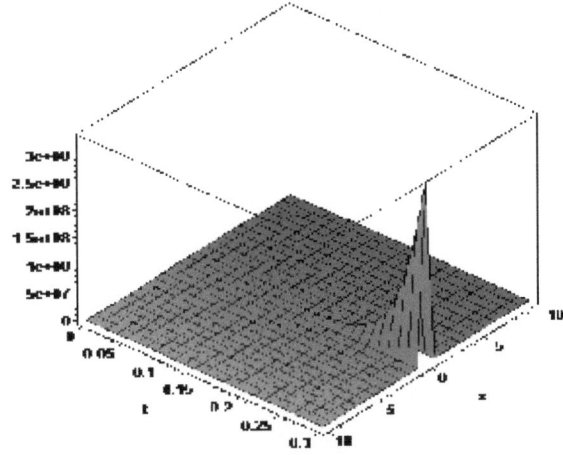

Approximation solution of Eq. (38-39) by ETDM.

5   Conclusion

The coupling of Adomian decomposition method (ADM) and Elzaki transform method proved very effective to solve nonlinear partial differential equations. The modified algorithm is suitable for such problems and is very userfriendly. From the obtained results, it is clear that the Elzaki transform decomposition method yields very accurate approximate solutions using only a few iterates, especially in the equations of the third order, where we note the emergence of the decomposition series solution after calculating the two first terms only. As a result, the conclusion that comes through this work is that (ETDM) can be applied to other nonlinear partial differential equations of higher order, due to the efficiency and flexibility in the application to get the possible results.

References

[1] G. Adomian, *Nonlinear Stochastic Systems Theory and Applications to Physics,* Kluwer Academic Publishers, Netherlands, 1989.

[2] G. Adomian and R. Rach, *Equality of partial solutions in the decomposition method for linear or nonlinear partial differential equations,* Comput. Math. Appl. 10, (1990), 9–12.

[3] G. Adomian, *Solving Frontier Problems of Physics: The Decomposition Method,* Kluwer Academic Publishers, Boston, 1994.

[4] G. Adomian, *Solution of physical problems by decomposition,* Comput. Math. Appl. 27, (1994), 145–154.

[5] G. Adomian, *Solutions of nonlinear P.D.E,* Appl. Math. Lett. 11, (1998), 121–123.

[6] J. H. He, *A new approach to nonlinear partial differential equations,* Comm. Nonlinear Sci. Numer. Simul., 2, (1997), 203-205.

[7] J. H. He, *Approximate analytical solution for seepage flow with fractional derivatives in porous media,* Comput. Meth. Appl. Mech. Eng., 167, (1998), 57-68.

[8] J. H. He, *A variational iteration approach to nonlinear problems and its applications,* Mech. Appl., 20, (1998), 30-31.

[9] J. H. He, *Variational iteration method for autonomous ordinary differential systems,* Appl. Math. Comput., 114, (2000), 115-123.

[10] J. H. He and X. H. Wu, *Variational iteration method: new development and applications,* Comput. Math. Appl., 54, (2007), 881-894.

[11] J. H. He, *Homotopy perturbation technique,* Comput. Meth. Appl. Mech. Eng., 178, (1999), 257–262.

[12] J. H. He, *Application of homotopy perturbation method to nonlinear wave equations,* Chaos Solitons Fractals, 26, (2005), 695–700.

[13] J. H. He, *A coupling method of homotopy technique and perturbation technique for nonlinear problems,* Int. J. Nonlinear Mech., 35, (2000), 37–43.

[14] J. H. He, *Some asymptotic methods for strongly nonlinear equations,* Int. J. Modern Phys., B, 20, (2006), 1141–1199.

[15] J. H. He, *A new perturbation technique which is also valid for large parameters,* J. Sound Vib., 229, (2000), 1257–1263.

[16] S. A. Khuri, *A Laplace decomposition algorithm applied to a class of nonlinear differential equations,* J. Math. Annl. Appl., 4, (2001), 141-155.

[17] S. A. Khuri, *A new approach to Bratus problem,* Appl. Math. Comput., 147, (2004), 31-136.

[18] D. Kumar, J. Singh and S. Rathore, *Sumudu Decomposition Method for Nonlinear Equations,* Int. Math. For., 7, (2012), 515 - 521.

[19] K. Afshan and M. Syed Tauseef, *Coupling of laplace transform and correction functional for wave equations,* W. J. Mod. Simul., 9, (2013), 173-180.

[20] A. S. Arife and A. Yildirim, *New Modified Variational Iteration Transform Method (MVITM) for solving eighth-order Boundary value problems in one Step,* W. Appl. Sci. J., 13, (2011), 2186 -2190.

[21] A. S. Abedl-Rady, S. Z. Rida, A. A. M. Arafa and H. R. Abedl-Rahim, *Variational Iteration Sumudu Transform Method for Solving Fractional Nonlinear Gas Dynamics Equation,* Int. J. Res. Stu. Sci. Eng. Tech., 1, (2014), 82-90.

[22] S. Kumara, A. Yildirim, Y. Khan and L. Weid, *A fractional model of the diffusion equation and its analytical solution using Laplace transform,* Scientia Iranica B 19, (2012), 1117-1123.

[23] J. Singh, D. Kumar and Sushila, *Homotopy Perturbation Sumudu Transform Method for Nonlinear Equations,* Adv. Theor. Appl. Mech., 4, (2011), 165 - 175.

[24] J. Singh, D. Kumar and Sushila, *Sumudu Homotopy Perturbation Technique,* Glo. J. Sci. Fron. Res., 11, (2011), 58-64.

[25] Sushila, J. Singh and Y. S. Shishodia, *An efficient analytical approach for MHD viscous flow over a stretching sheet via homotopy perturbation sumudu transform method,* Ain Shams Eng. J., 4, (2013), 549–555.

[26] T. M. Elzaki and S. M. Elzaki and E. A. Elnour, *On the New Integral Transform "ELzaki Transform" Fundamental Properties Investigations and Applications,* Glo. J. Math. Sci., 4, (2012), 1-13.

[27] T. M. Elzaki and E. M. A. Hilal, *Homotopy Perturbation and Elzaki Transform for Solving Nonlinear Partial Differential Equations,* Math. Theor. Mod., 2, (2012), 33-42.

[28] E. M. Abd Elmohmoud and T. M. Elzaki, *Elzaki Transform of Derivative Expressed by Heaviside Function,* W. Appl. Sci. J., 32, (2014), 1686-1689.

[29] M. Khalid, M. Sultana, F. Zaidi and U. Arshad, *An Elzaki Transform Decomposition Algorithm Applied to a Class of Non-Linear Differential Equations,* J. Natural Sci. Res., 5, (2015), 48-56.

[30] Y. Zhu, Q. Chang and S. Wu, *A new algorithm for calculating Adomian polynomials,* Appl. Math. Comput., 169, (2005), 402–416.

[31] M. M. Hosseini and H. Nasabzadeh, *On the convergence of Adomian decomposition method,* Appl. Math. Comput., 182, (2006), 536–543.

[32] A. M. Wazwaz, *The variational iteration method for rational solutions for KdV, K(2,2), Burgers, and cubic Boussinesq equations,* J. Comp. Appl. Math., 207, (2007), 18 – 23.

Testing an Algorithm for an Adaptive Wavelet Based Method in Mathematica and Comparing the Method with NDSolve

M. Schuchmann and M. Rasguljajew from the Darmstadt University of Applied Sciences

Abstract

M. Schuchmann developed in 2013 and 2014 an adaptive wavelet based algorithm to numerically solve differential equations. This algorithm can be applied to ODEs with various orders and even to PDEs and differential-algebraic equations (DAEs). In this Paper we compare the solutions of boundary value problems of second order ODEs calculated by the adaptive wavelet algorithm with the solution of the Mathematica function NDSolve. In NDSolve we have different explicit or implicit integration methods (like implicit or explicit Runge-Kutta, Adams or BDF method) and it will automatically reduce the step size or change the method depending on the problem, so NDSolve can handle stiff and non-stiff problems.

We have implemented the algorithm in a Mathematica module and applied several tests. The used wavelet collocation method is costly, like collocation method are costly in general. But for stiff or unstable problems it has an advantage and for these cases explicit methods cannot be used. We used the Shannon wavelet in our module, but it can be modified for other wavelets and the ODE can be given in implicit form.

Introduction

Our module can be used for ODEs of order 1 and 2. It can easily be adjusted for higher order ODEs. It uses an adaptive algorithm to adjust the parameters of the wavelet based method.

The method uses an approximation function y_j constructed with a subset of the bases of V_j. With that approximation function we minimize the sum of squares of the residuals at so called collocation points t_i. For a certain number of collocation points this method is equivalent to the collocation method using that 'wavelet bases'. The initial points or boundary values are integrated in the minimization process, but they could also be used as restraints.

Here we have to adjust several parameters like the solution index j, the number of bases functions out of V_j and the number and the position of the collocation points t_i. The adjustment of these parameters is the objective of the adaptive algorithm. For the construction of the algorithm theoretical investigations have been necessary and M. Schuchmann has used an error estimation (see [16]) which is based on the criteria Q_a (see (2)) and he also evaluated many simulation to use the information of the minimum of Q (see (1)) and Q_a in order to adjust the parameters.

The advantage of the wavelet collocation method is that like other collocation method it can also be applied to stiff differential equations. Moreover, it can even be used for non-stable problems and even for DAEs. As an approximation we not only get points but an approximation function. Compared to collocation methods, for example, based on polynomials (see [3]), we get one approximation function for the whole approximation Interval and it can also cover a larger interval. The method can also be used for implicit ODEs.

These are the advantages of a wavelet collocation method as well as using the approximation for extrapolation outside the original approximation interval. A disadvantage is that these methods, like many boundary value methods or implicit methods, are very costly. But the method is robust an it can applied to stiff or instable systems as well to implicit ODEs or DAEs.

Theory

In the wavelet theory a scaling function ϕ is used, which belongs to a MSA (multi scale analysis). From the MSA we know, that we can construct an orthonormal basis of a closed subspace V_j, where V_j belongs to a the sequence of subspaces with the following property:

$$ ... \subset V_{-1} \subset V_0 \subset V_1 \subset ... \subset L^2(R), $$

$\{\phi_{j,k}(t)\}_{k \in Z}$ is an orthonormal basis of V_j with $\phi_{j,k}(t) = 2^{j/2}\phi(2^j t - k)$.

We use the following approximation function

$$ y_j(t) := \sum_{k=k_{min}}^{k_{max}} c_k \cdot \phi_{j,k}(t) \quad , \text{ with } \phi \in C^s(R). $$

k_{max} and k_{min} depend on the approximation interval $[t_0, t_{end}]$ (see [14]). s is the order of the ODE.

Now we can approximate the solution of an initial value problem $y' = f(y,t)$ and $y(t_0) = y_0$ by minimization of the following function

$$ (1) \qquad Q(c) = \sum_{i=1}^{m} \left\| y_j'(t_i) - f(y_j(t_i), t_i) \right\|_2^2 + \left\| y_j(t_0) - y_0 \right\|_2^2 . $$

For $m = |k_{max} - k_{min}|$ we get an equivalent problem:

$$ y_j'(t_i) = f(y_j(t_i), t_i) \text{ for } i = 1, 2, ..., m \text{ and } y_j(t_0) = y_0. $$

This is the classical collocation method, whereas the minimum residual method is more general.

We will use equidistant points or collocation points t_i with $t_i = t_0 + i \cdot h$ and

$$ h = \frac{t_{end} - t_0}{m} \qquad . $$

To detect large residuals in other places than the collocation points, we have a further value used for comparison with Q_{min} (here in y_j the vector c will be set to the value in the minimum of Q, see (1)).

$$ (2) \qquad Q_a = \sum_{i=1}^{m_a} \left\| y_j'(\tau_i) - f(y_j(\tau_i), \tau_i) \right\|_2^2 + \left\| y_j(t_0) - y_0 \right\|_2^2 $$

with $\tau_i = t_0 + i \cdot h/a$. $m_a = a \cdot m$ with $a > 1$ as an integer. Since the wavelet collocation method provides a whole approximation function y_j and not only points, we can calculate Q_a without additional effort. If $Q_a \gg Q_{min}$ (and Q_{min} was very small) then m (the number of collocation points) should be increased. When comparing Q_{min} with Q_a, Q_a should be weighted by $1/a$ if a is large. In the simulations $a = 2$ proved sufficient.

Q_a can additionally be justified by an error estimation of the residuals at theoretically any number of points. This was derived by M. Schuchmann (see [16]). In this error estimate a certain value occurs as a factor. Q_a represents the Riemann sum for this value i.e. this can be approximated by Q_a.

If we have a second Order ODE

$$F(y'', y', y, t) = 0$$

with boundary conditions

$$y(t_0) = y_0$$
$$\text{and}$$
$$y(t_{end}) = y_{end}$$

In the following example, we minimize

$$Q(c) = \sum_{i=1}^{m} \left\| F(y_j''(t_i), y_j'(t_i), y_j(t_i), t_i) \right\|_2^2 + \left\| y_j(t_0) - y_0 \right\|_2^2 + \left\| y_j(t_{end}) - y_{end} \right\|_2^2 .$$

Analogous we treat conditions of the form

$$y(t_0) = y_0$$
$$\text{and}$$
$$y'(t_0) = y'_0$$

t_0 in the conditions (boundary or initial value) must not be the same as t_0 from the approximation interval, that means we can even address the conditions $y(\tilde{t}_0) = y_0$ and $y'(\tilde{t}_0) = y'_0$ with $\tilde{t}_0 \neq t_0$.

The Modules

We have two modules: one for the wavelet collocation method (WCollocationS2) and one for the algorithm (WCollocationS2Alg). The first module is for the minimization of Q with certain j, k_{max} and r. The second module applies the algorithm to calculate the 'optimal' j, k_{max} and r. Here we can solve numerically first order ODEs with initial values, where the ODE can be given implicit (or explicit). The module needs only the left side of the equation

$$f(y'(t), y(t), t) = 0 \text{ or } y'(t) - f(y(t), t) = 0.$$

Analogous we can solve numerically ODEs of second order with initial values or boundary values.

In the module for the algorithm (`WCollocationS2Alg`) we have implemented the following algorithm:

We only need specify $k_{max}^{(0)}$ (as a positive integer). k_{min} will then be adjusted for the approximation interval. If the approximation interval $I = [t_0, t_{end}]$ is symmetric around 0, k_{min} is set to $-k_{max}^{(0)}$ (and $k_{max} = k_{max}^{(0)}$). In the other case the module calculates

$$k_0 = \text{round}(2^j(t_0 + t_{end})/2)$$

and sets $k_{max} = k_{max}^{(0)} + k_0$ and $k_{min} = -k_{max}^{(0)} + k_0$, where $k_{max}^{(0)}$ can be changed during the iteration (see algorithm at the following page).
$k_{max}^{(0)}$ is set to be at least 10.

The step size is calculated as follows:

$$h = \frac{t_{end} - t_0}{m} \quad \text{with} \quad m = r \cdot |k_{max}^{(0)}| \, .$$

The initial r can be set in the module. It should be set to a value of at least 2 (for first order ODEs we have with $r = 2$ the classical collocation case), but the module accepts 1, too. $k_{max}^{(0)}$, j and r are positive integers.

Suitable positive real numbers ϵ_1 and ϵ_2 should be chosen (if it does not get specified, the module choose $\epsilon_1 = 10^{-6}$ and $\epsilon_2 = 10^{-2}$). The default value of a is 2, of $k_{max}^{(0)}$ is 15 and j is 1.

The implemented algorithm:
The algorithm calculates Q_{min} and Q_a.

If $Q_{min} \leq \epsilon_1$, then it is checked whether $Q_a \leq \epsilon_2$ applies (with $a > 1$, for example $a = 2$). If both conditions are met, then the iteration is finished.

Generally we suggested the following algorithm: If $Q_{min} \leq \epsilon_1$ is not met, j is incremented by 1 (if a sufficient number of basis functions $\phi_{j,k}$ are chosen with respect to the approximation interval I). If $Q_{min} < \epsilon_1$ is met but $Q_2 < \epsilon_2$ not, then m should be increased.

The module works with maximum 7 steps:
In the first step, if $Q_{min} > \epsilon_1$ then $k_{max}^{(0)}$ is increased by 5 if $k_{max}^{(0)} \leq 40$ and j is increased by 1 if $k_{max}^{(0)} > 40$. If $Q_a > \epsilon_2$ and $Q_{min} \leq \epsilon_1$ then r is increased by 1.

After the first step, 5 steps follow.
If $Q_{min} > \epsilon_1$ then $k_{max}^{(0)}$ is increased by 5 (if $k_{max}^{(0)} < 45$), j by 1 and r by 1 (if $r < 9$).
If $Q_a > \epsilon_2$ and $Q_{min} \leq \epsilon_1$ then r is increased by 1. When j is incremented by 1 we could double k_{max}, but this can lead to very big minimization problems (because the length of c would double). Many simulations showed, that with k_{max} must not be so large as a $L^2(R)$ approximation needs (see example iv).

If after all these steps $Q_{min} > \epsilon_1$ or $Q_a > \epsilon_2$, then Q will be minimized for the last time.

The step counter in the mathematica module for the algorithm (WCollocationS2) ist called z ($z = 0$ to $z = zmax$, $zmax = 4$) and the maximum number of k_{max} is set to 45. The maximum number of r ist 9. In example iv we needed a biger k_{max} and here was set the upper bound of k_{max} to 100.

The Module for the algorithm can be called with

```
WCollocationS2Alg[ODE,t₀,y₀,t₀',t_end',k_max⁽⁰⁾(15),j(1),r(2),a(2),
ε₁(10⁻⁶),ε₂(10⁻²)]
```

and (in brackets the default values) the collocation can be called with:

```
WCollocationS2[ODE,t₀,y₀,t₀',t_end',k_max⁽⁰⁾ (15),j(1),r(2),a(2)]
```

The first argument is the left side of the ODE in the form $F(y', y, t) = 0$ or $F(y'', y', y, t) = 0$ in Mathematica notation, that means y is y[t], y' is y'[t]. t_0 is the initial value or a List in the form $\{t_0, t_{end}\}$ an analogous y(t_0) or $\{y(t_0), y(t_{end})\}$. If we have a initial value problem for an second order ODE, then we can specify t_0 and $\{y(t_0), y'(t_0)\}$. The approximation interval boundaries are here called t_0' and t_{end}' (must not be identically with the initial or boundary t_0). The next parameters are optional: j, r, a (for Q_a). For the algorithm module we have additional ϵ_1 and ϵ_2. Because of $Q_{min} \leq Q_a$ (for $a = 2, 3, ...$) or in most practical cases $Q_{min} < Q_a$ the module gives a warning if $\epsilon_1 > \epsilon_2$ and sets $\epsilon_2 = 10\epsilon_1$. If you need very good approximations you may choose a smaller ϵ_1 and ϵ_2, but ϵ_2 should not be too small for problems with large slopes or large curvatures (for example for stiff problems). The modules can be downloaded from http://algorithm.jatam.de/.

For the minimization the Mathematica function FindMinimum will be used. Here we can get the following comment from Mathematica:
FindMinimum::sszero: The step size in the search has become less than the tolerance prescribed by the PrecisionGoal option, but the gradient is larger than the tolerance specified by the AccuracyGoal option. There is a possibility that the method has stalled at a point that is not a local minimum.

We got this comment in almost all iterations during the first iteration steps, but without any bad consequences for the approximation.

The algorithm can be also applied on higher order ODEs or (according to a generalization) on PDEs.

The second order ODEs have been found on the website of Jeff Cash (Imperial College, London). In the tests we used Mathematica 9.0 and 10.3.

The modules can be downloaded over the address http://jatam.de/algorithm/Wavelet-Approximation-Algorithm.nb.

Example I

We apply the algorithm on a second order ODE with boundary conditions:

$y'' - y'/\mu = 0$ with $\mu = 1/100$ and with $y(0) = 1$, $y(1) = 0$, the approximation interval is $I = [0, 1]$.

We called the module for the algorithm with:

```
WCollocationS2Alg[y''[t] - 100 y'[t],{0,1},{1,0},0.,1.,15,1,2,2]
```

Mathematica NDSolve has problems:
NDSolve::bvluc: The equations derived from the boundary conditions are numerically ill-conditioned. The boundary conditions may not be sufficient to uniquely define a solution. The computed solution may match the boundary conditions poorly. »

Here is the graph of η - y (η is the NDSolve solution), were we can see, that the numerical solution of NDSolve has big deviations:

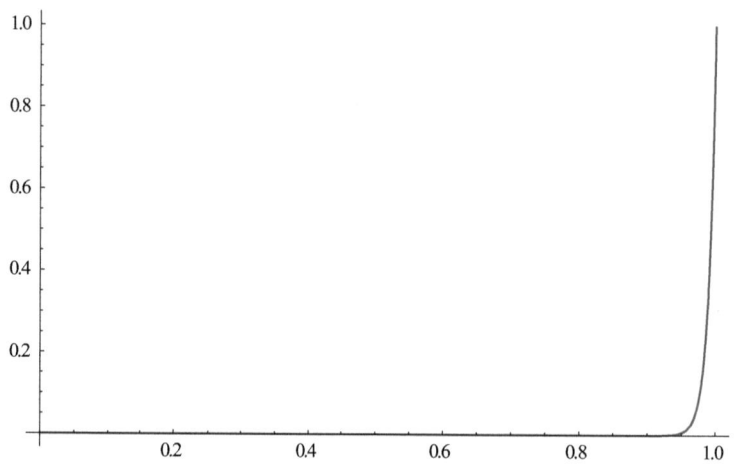

Here is the graph of η.

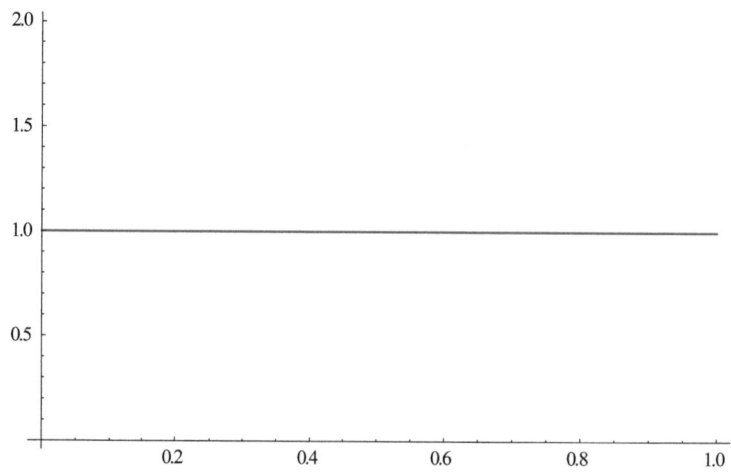

Now we see the iteration-protocol:

$k_{max}^{(0)}$	j	r	Q_{min}	Q_a
15	1	2	0.499997	0.50022
20	1	2	0.499999	0.500003
25	2	3	0.499996	0.5
30	3	4	0.496216	0.513244
35	4	5	0.0074078	0.0520069
40	5	6	3.88801×10^{-8}	4.03185×10^{-6}

For critical examples we could start with a higher k_{max}, j and r.

Here are the graphs of y_5 and y (we see no differences graphically):

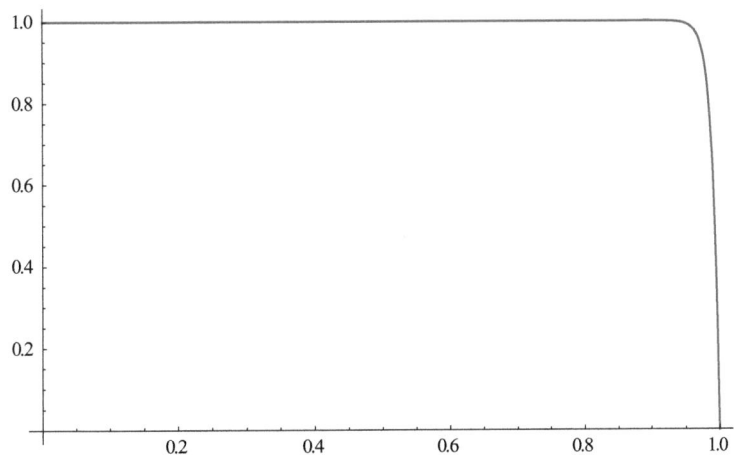

Here is the graph of y_5 - y:

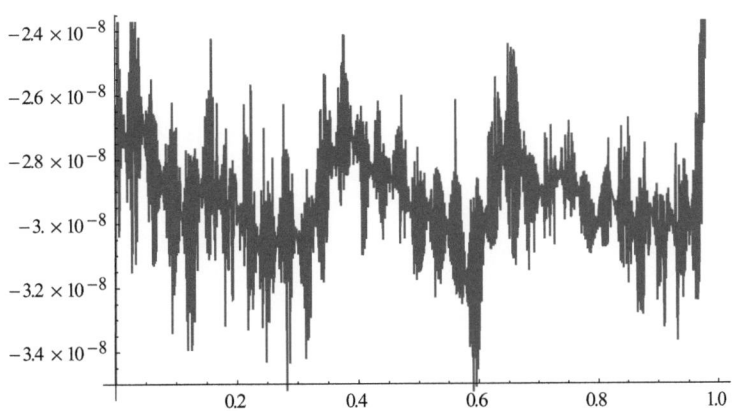

Example II

We apply the algorithm on a second order ODE with boundary conditions:

$y'' = (-y' + (1+\mu) \cdot y)/\mu$ with $\mu = 1/50$ and with $y(-1) = 1+e^{-2}$, $y(1) = 1+e^{-2(1+\mu)/\mu}$, the approximation interval is $I = [-1, 1]$.

We called the module for the algorithm with:

```
WCollocationS2Alg[-51y[t]+50y'[t]+y''[t],{-1,1},{1+e⁻², 1+e⁻¹⁰²},-1.,1.,30,
3,2,2]
```

Here is the graph of $\eta - y$ (η is the NDSolve solution), were we can see, that the numerical solution of NDSolve has no big deviations:

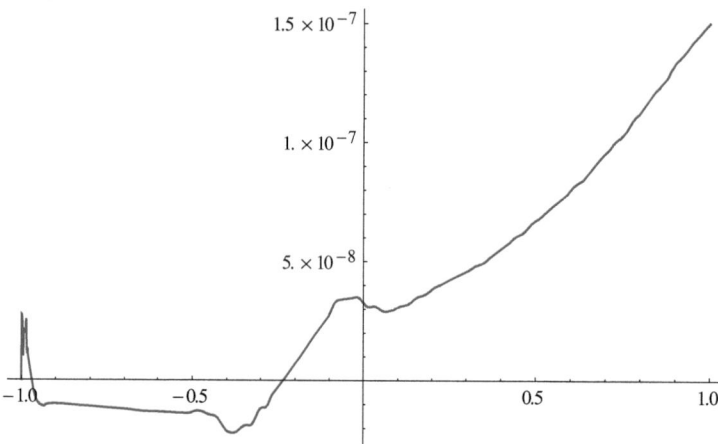

Here is the graph of η.

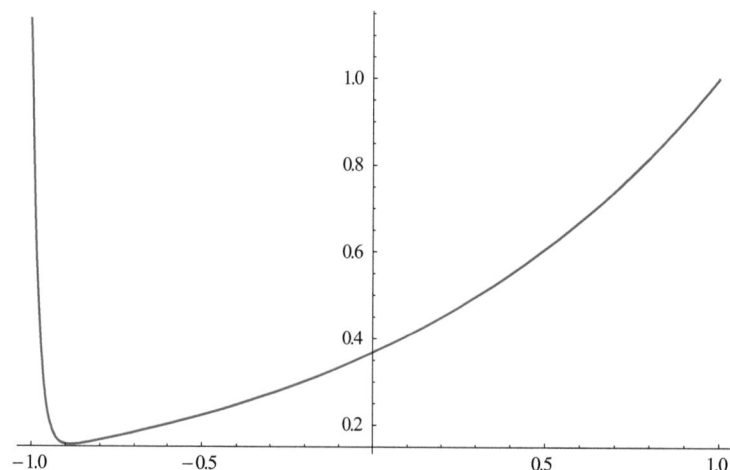

Now we see the iteration-protocol:

$k_{max}^{(0)}$	j	r	Q_{min}	Q_a
30	3	2	6.93893×10^{-9}	18.6674
30	3	3	8.46827×10^{-6}	1.18102
35	4	4	3.44103×10^{-10}	0.000228873

For critical examples we could start with a higher k_{max}, j and r. The solution of the first step (with $j = 3$ and $r = 2$) is not bad, the big Q_a of 18.6674 results form a big residual value at one point ($t = -59/60$).

Here are the graphs of y_4 and y, where we see a good approximation, too.

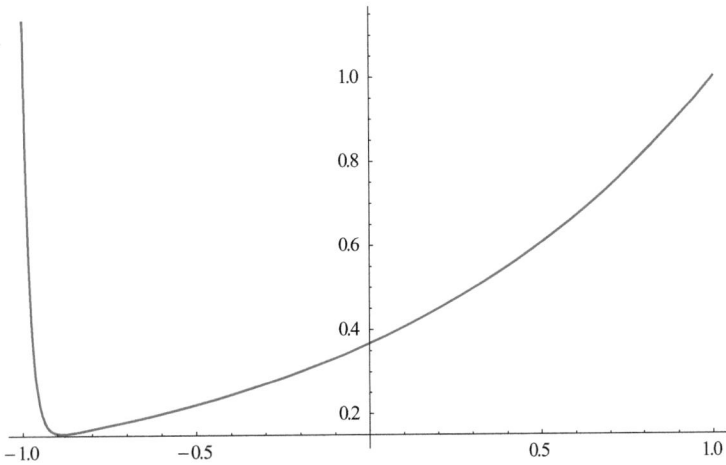

Here is the graph of $y_4 - y$:

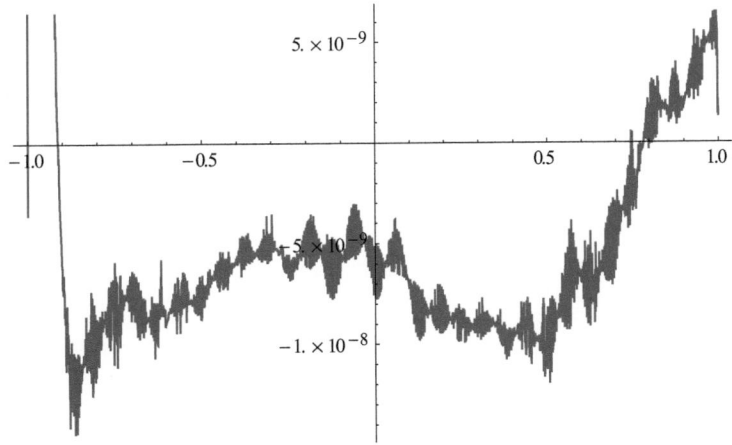

Example III

We apply the algorithm on a second order ODE with boundary conditions:

$y'' = (-t \cdot y' - \mu \cdot \pi^2 \cdot cos(\pi t) - \pi t \, sin(\pi t)) / \mu$ with $\mu = 1/10$ and with $y(-1) = -2$, $y(1) = 0$, the approximation interval is $I = [-1, 1]$.

We called the module for the algorithm with:

```
WCollocationS2Alg[π² Cos[π t]+10 π t Sin[π t]+10 t y'[t]+y''[t],{-1,1},{-2,
0},-1.,1.,15,1,2,2]
```

Here is the graph of $\eta - y$ (η is the NDSolve solution):

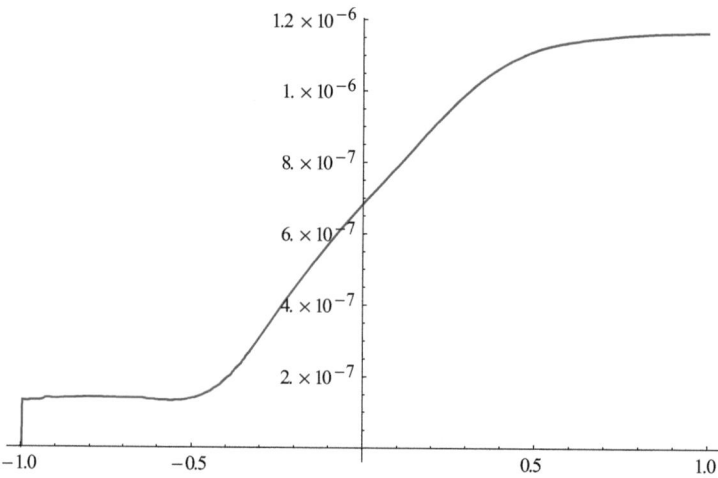

Here is the graph of η.

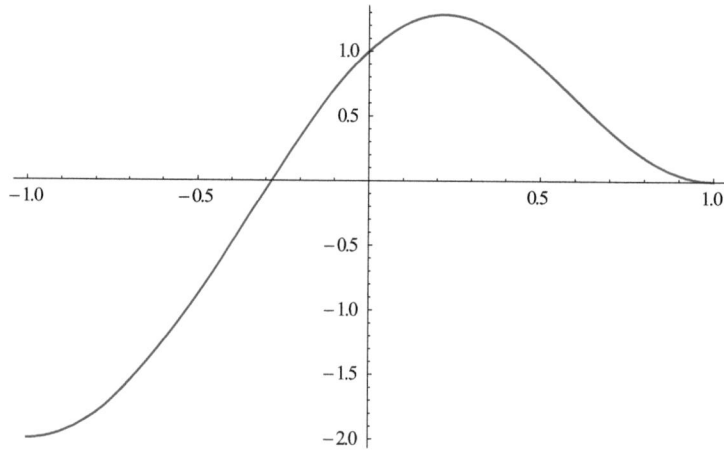

Now we see the iteration-protocol:

$k_{max}^{(0)}$	j	r	Q_{min}	Q_a
15	1	2	5.8001×10^{-13}	1.85559×10^{-7}

For critical examples we could start with a higher k_{max}, j and r.

Here are the graphs of y_1 and y (we see no differences):

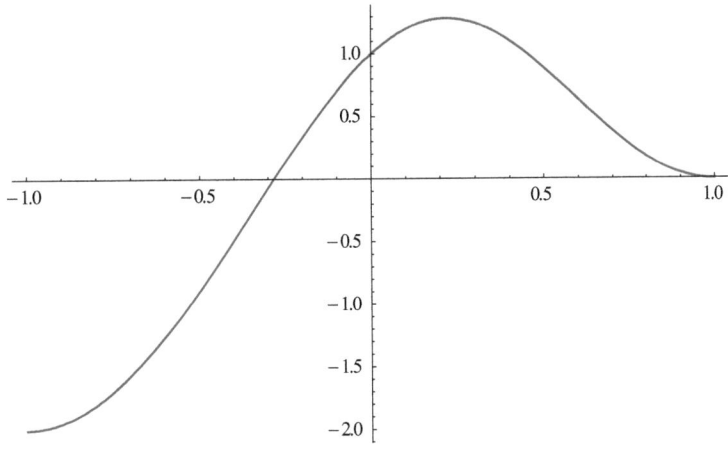

Here is the graph of $y_1 - y$:

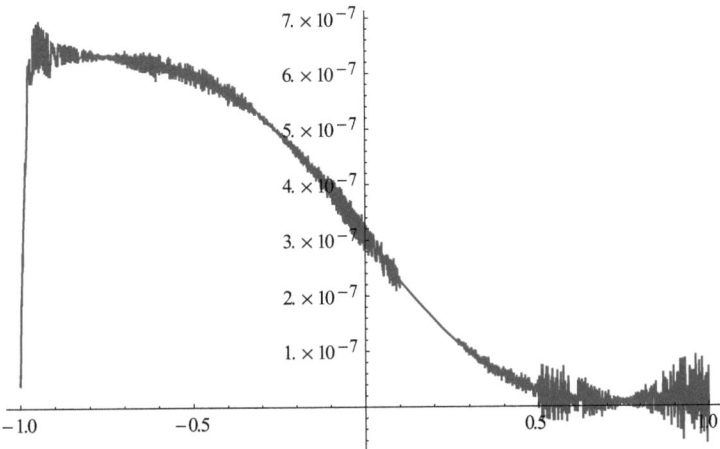

At last we see graphically the relation between a und Q_a/a in this example:

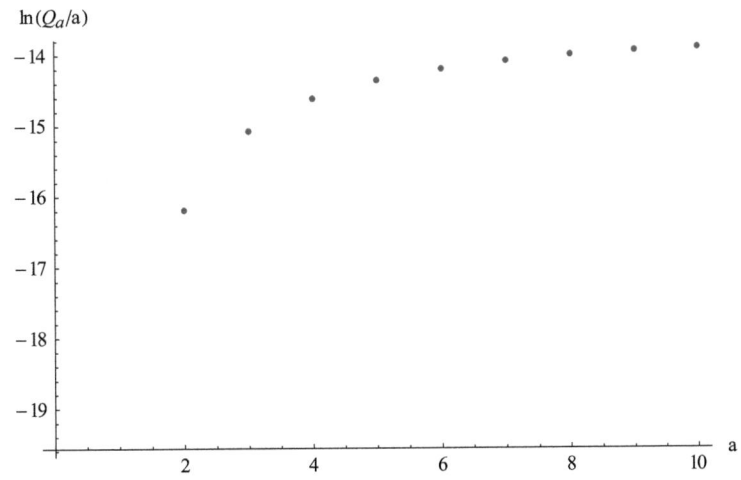

When we set $\mu = 1/100$, then NDSolve get problems:

```
NDSolve::bvluc: The equations derived from the boundary conditions are numerically
ill-conditioned. The boundary conditions may not be sufficient to uniquely define a
solution. The computed solution may match the boundary conditions poorly. ≫
```

NDSolve::berr: There are significant errors _{0.,-454021.}_ in the boundary value residuals. Returning the best solution found. ≫

Here we get a very bad approximation:
η - y (η is the NDSolve solution):

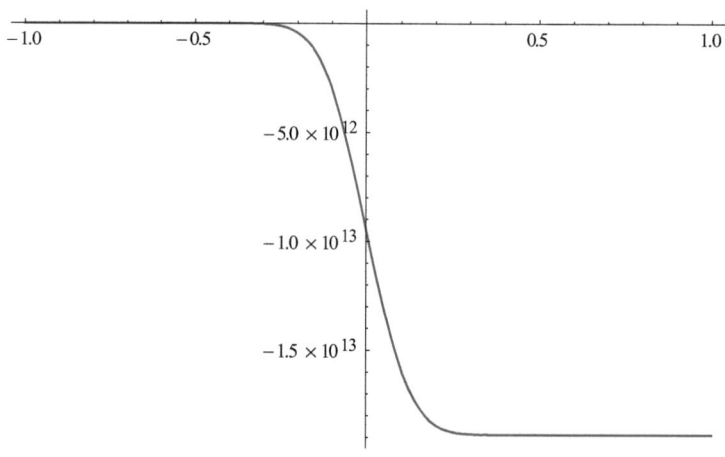

The algorithm has no problems:

```
WCollocationS2Alg[π² Cos[π t]+100 π t Sin[π t]+100 t y'[t]+y''[t],{-1,1},{-2,
0},-1.,1.,15,1,2,2]
```

The graphs of y_4 and y:

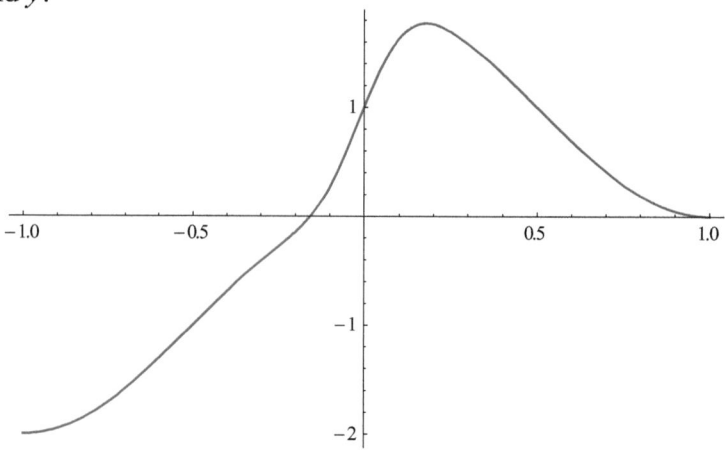

The graphs of y_4 - y:

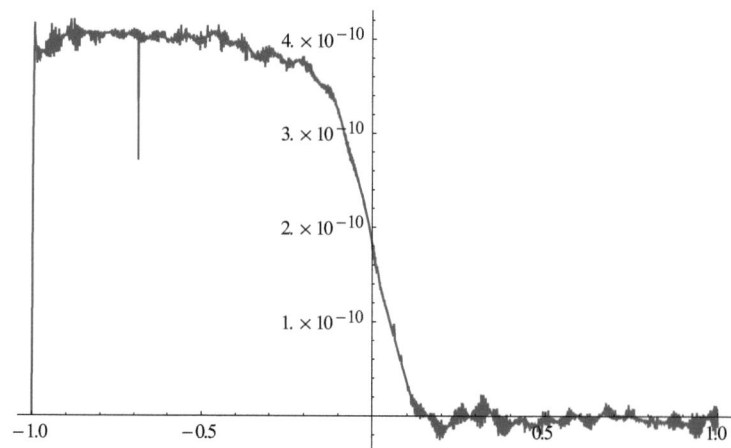

At last the iteration protocol:

$k_{max}^{(0)}$	j	r	Q_{min}	Q_a
15	1	2	1.44579	319690.
20	1	2	1.97176	258.584
25	2	3	1.66245	195.992
30	3	4	0.00284779	5.50948
35	4	5	4.75163×10^{-14}	1.06944×10^{-10}

Example IV

In this example we will see, that the maximum number of k_{max} in the module (the maximum value of *kmax* was set to 45) should be larger in problems, which needs a large j. We apply the algorithm on a second order ODE with boundary conditions:

$y'' = (-4t\,y' - 2y)/(\mu + t^2)$ with $\mu = 1/50$ and with $y(-1) = 1/(1 + \mu)$, $y(1) = 1/(1 + \mu)$, the approximation interval is $I = [-1, 1]$.

We called the module for the algorithm with:

```
WCollocationS2Alg[-(-2y[t]- 4t*y'[t])/(1/50+t²)+y''[t],{-1,1},{10/11,10/11},
-1.,1.,30,5,3,2]
```

Mathematica NDSolve has no problems:

Here is the graph of $\eta - y$ (η is the NDSolve solution):

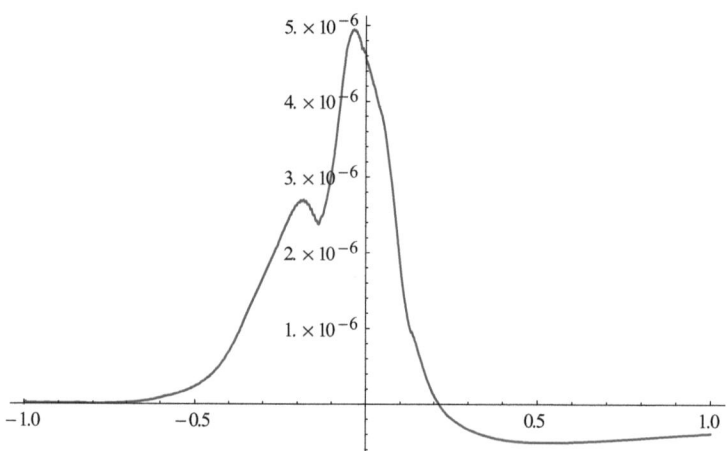

Here is the graph of η.

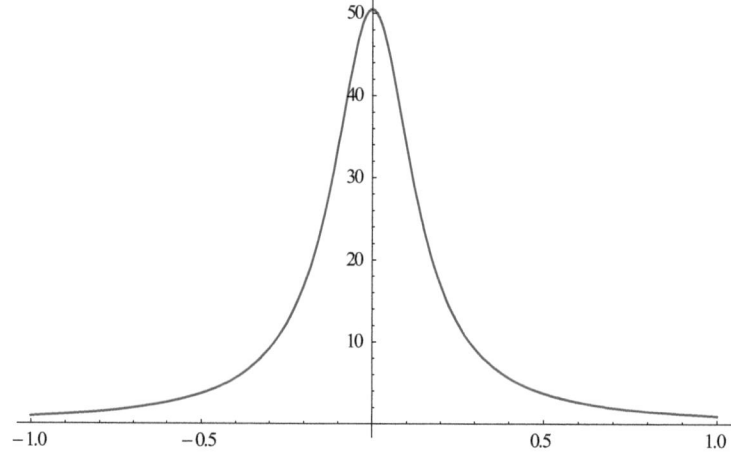

With the used starting values the algorithm stops after the maximum number of steps has been made and the warning came that the solution does not satisfy the convergence criteria $Q_{min} \leq \epsilon_1$ and $Q_a \leq \epsilon_2$. So we know that the solution is not usable.

Now we see the iteration-protocol:

$k_{max}^{(0)}$	j	r	Q_{min}	Q_a
30	5	3	1.96059	1.96059
35	5	3	0.96076	22.8715
40	6	4	1.96059	1.96059
45	7	5	1.96059	1.9606
45	8	6	1.96059	1.96115
45	9	7	1.96059	1.96099
45	10	8	1.96059	2.49262

The module prints: `Warning: Q_min or Q_a is bigger than the tolerance!`

We see in the iteration protocol, that the module has a maximum number of k_{max}. For critical problems, where we need a bigger j, the maximum number of k_{max} should be set to a higher value than 45 in `WCollocationS2Alg`. With $k_{max}^{(0)}$ less than 2^j the method cannot get a solution (with a small Q_{min}, because Q_{min} is in that case $\geq y(0)^2 + y(1)^2$) with the Shannon ϕ with that boundary conditions, because at $t_{end} = 1$ we get the boundary condition

$$y_j(1) := \sum_{k=k_{min}}^{k_{max}} c_k \cdot \phi_{j,k}(1) = \sum_{k=k_{min}}^{k_{max}} c_k \cdot 2^{j/2} \phi(2^j \cdot 1 - k) \overset{!}{=} y(1)$$

and $\phi(m) = 0$ for integer $m \neq 0$ and $\phi(1) = 1$. So if k_{max} is less than 2^j the boundary condition cannot be fulfilled if $y(1) \neq 0$. Because if $y(-1) \neq 0$ we get the same for k_{min}. So with that boundary conditions we get $k_{min} \leq -2^j$ and $k_{max} \geq 2^j$. Otherwise $y_j(\pm 1) = 0$. With integer values of the boundaries t_0 and t_{end} general k_{max} should be greater or equal $2^j t_{end}$. Because of k_{min} should be less or equal $2^j t_0$, in the module $k_{max}^{(0)}$ should be greater or equal (only if the expression is integer) $(2^j t_{end} - 2^j t_0)/2$, because in the module $k_{max}^{(0)}$ is positive (the module shifts automatically the summation area, $k_{max} = k_{max}^{(0)} + k_0$ and $k_{min} = -k_{max}^{(0)} + k_0$).

When set *kmaxmax* = 100 an apply the method with

```
WCollocationS2Alg[-(-2y[t]- 4t*y'[t])/(1/50+t^2)+y''[t],{-1,1},{10/11,10/11},
-1.,1.,80,6,8,2]
```

then the algorithm stops directly with $k_{max} = 80$, $j = 6$, $r = 8$, $Q_{min} = 1.88584 \times 10^{-12}$ and $Q_a = 9.84522 \times 10^{-10}$. Here are the graphs of $y_6 - y$:

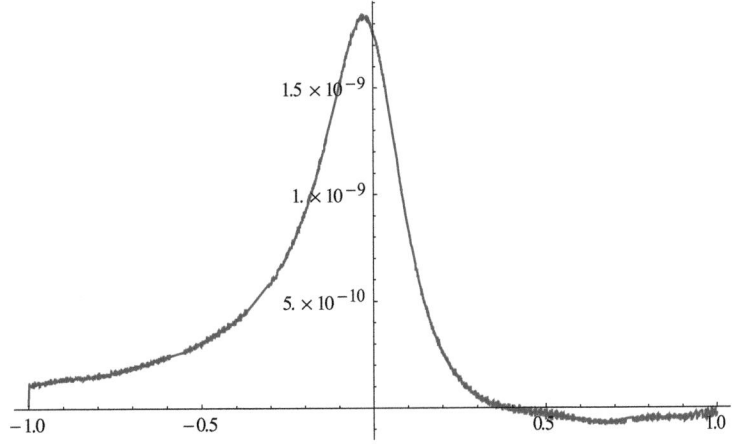

A direct approximation makes no problems, too. For example, with

```
WCollocationS2Alg[y[t]-fe[t],-1,50/51,-1.,1.,25,2,4,2]
Fe[t_]:=50/(1+50t²)
```

the algorithm stops with $k_{max} = 45, j = 5, r = 7$, $Q_{min} = 5.83423 \times 10^{-10}$ and $Q_a = 4.52406 \times 10^{-9}$.

Here are the graphs of y_5 - y:

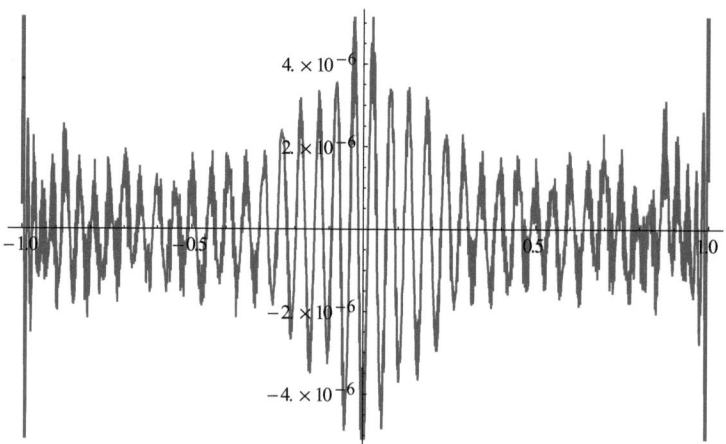

The difference between the second derivations is relative large (graph of $y_5'' - y''$):

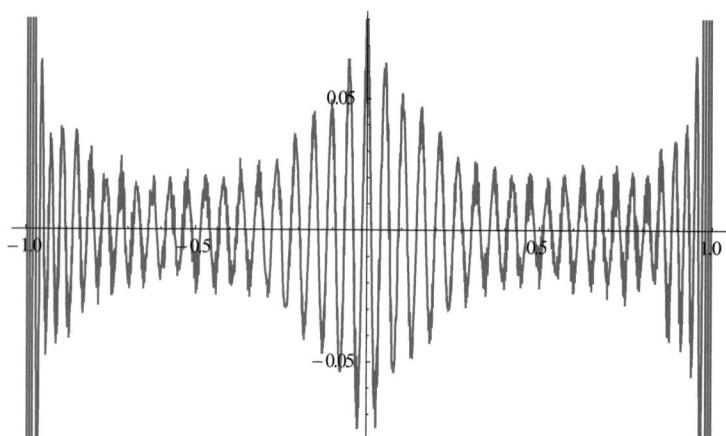

Here we get:

$$\sum_{i=0}^{m} (y_j(t_i) - y(t_i))^2 = 5.83423 \times 10^{-10}$$

with $\qquad h = \dfrac{t_{end} - t_0}{m}$ $\qquad$ with $m = r \cdot |k_{max}^{(0)}| = 45 \cdot 7$, $t_0 = -1$ and $t_{end} = 1$.

The derivation of the second order derivatives is much larger:

$$\sum_{i=1}^{m} (y_j''(t_i) - y''(t_i))^2 = 8308.59$$

The biggest difference we get at the beginning and at the end of the approximation interval with 3970.8 and 4307.34.

Comparing the L^2 approximation with the direct approximation on the interval [-1,1]:
We can not apply the information about the $L^2(R)$ approximation $\tilde{y}_j$ from y on

$$\underbrace{span\ \{\phi_{j,k}\}_{k=k_{min},k_{min}+1,...,k_{max}}}_{=:S_j} \subset V_j$$

to get the right k_{min} and k_{max} for the algorithm, because the $L^2(R)$ approximation may need a lot bigger k_{max} than the direct approximation through (4) on the interval [-1, 1] (to get nearly the same quality of approximation), like in our example, where the decay of the coefficients $\tilde{c}_k$ is very poor.

For the $L^2(R)$ approximation the coefficients will be calculated as usual with orthogonal bases (we assume for easier notation, that the scaling function and y is real valued):

$$\tilde{c}_k = \left\langle y,\phi_{j,k} \right\rangle_{L^2(R)} = \int_R y(t)\cdot\phi_{j,k}(t)dt$$

And so we get the orthogonal projection from y on S_j:

$$\tilde{y}_j(t) := \sum_{k=k_{min}}^{k_{max}} \tilde{c}_k \cdot \phi_{j,k}(t)$$

Generally $\left\|\tilde{y}_j - y\right\|_{L^2(I)} \geq \left\|\hat{y}_j - y\right\|_{L^2(I)}$ with $I \subset R$. Here $\tilde{y}_j$ is the best approximation on R, calculated through

$$min\left\|y_j - y\right\|_{L^2(R)} = \left\|\tilde{y}_j - y\right\|_{L^2(R)}$$

and $\hat{y}_j$ is the best approximation on I, calculated through

$$(3)\quad min\left\|y_j - y\right\|_{L^2(I)} = \left\|\hat{y}_j - y\right\|_{L^2(I)}.$$

The reason for that is because $\tilde{y}_j$ is the best approximation according to the $L^2(R)$ norm on R but $\hat{y}_j$ is the best approximation from y only on the interval I as a part of R. The direct approximation is the numerical solution of the minimum problem (3) above, so y_j is the numerical approximation of $\hat{y}_j$ or the solution of (4). Here - for easier notation - we named the solution of the minimum problems the same as the unknown functions.

Theoretically we would get the solution of the continuous minimum problem (3) through the following considerations:

We calculate instead of (3) the orthogonal projection of a function $\breve{y}$ on S_j. The function $\breve{y}$ is on I identical to y and on $R \setminus I$ identically to our function y_j of S_j as a part of V_j. So $\breve{y} = y_I + y_{R\setminus I}$ (where y_I vanishes on R\I and $y_{R\setminus I}$ vanishes on I):

$$\breve{y}(t) = 1_I(t) \cdot y(t) + 1_{R\setminus I}(t) \cdot \sum_{k=k_{min}}^{k_{max}} c_k \cdot \phi_{j,k}(t) \text{ with indicator function 1.}$$

So we approximate y only on I. Outside I the approximation function has no restricts. An other and a worse approximation we would get through the orthogonal projection from $1_I \cdot y$ on V_j. The reason is that we would cut the function y and this would lead generally to a bad decay behavior in the Fourier space, see [19].

Here we get the coefficients c_k through:

$$c_k = \left\langle \breve{y}, \phi_{j,k} \right\rangle_{L^2(R)} = \left\langle y, \phi_{j,k} \right\rangle_{L^2(I)} + \left\langle y_j, \phi_{j,k} \right\rangle_{L^2(R\setminus I)} = \left\langle y, \phi_{j,k} \right\rangle_{L^2(I)} + \sum_{k=k_{min}}^{k_{max}} c_l \cdot \left\langle \phi_{l,k}, \phi_{j,k} \right\rangle_{L^2(R\setminus I)}$$

For I = R we would get the best approximation on R through $\left\langle y, \phi_{j,k} \right\rangle_{L^2(R)} \cdot \{\phi_{j,k}\}_k$ is general no orthogonal system on $L^2(R \setminus I)$ (only if the support of $\phi_{j,k}$ is in $R \setminus I$). So we get:

$$c_k - \sum_{k=k_{min}}^{k_{max}} c_l \cdot \left\langle \phi_{l,k}, \phi_{j,k} \right\rangle_{L^2(R\setminus I)} = \left\langle y, \phi_{j,k} \right\rangle_{L^2(I)}$$

$$\sum_{k=k_{min}}^{k_{max}} c_l \cdot \delta_{l,k} - c_l \cdot \left\langle \phi_{l,k}, \phi_{j,k} \right\rangle_{L^2(R\setminus I)} = \left\langle y, \phi_{j,k} \right\rangle_{L^2(I)}$$

$$\sum_{k=k_{min}}^{k_{max}} c_l \cdot \left\langle \phi_{l,k}, \phi_{j,k} \right\rangle_{L^2(R)} - c_l \cdot \left\langle \phi_{l,k}, \phi_{j,k} \right\rangle_{L^2(R\setminus I)} = \left\langle y, \phi_{j,k} \right\rangle_{L^2(I)}$$

$$\sum_{k=k_{min}}^{k_{max}} c_l \cdot \underbrace{\left\langle \phi_{l,k}, \phi_{j,k} \right\rangle_{L^2(I)}}_{:=a_{l,k}} = \underbrace{\left\langle y, \phi_{j,k} \right\rangle_{L^2(I)}}_{:=u_k} \quad , \text{ for } l = k_{min}, \ldots, k_{max} \quad (5)$$

That is the normal equation for the vector c: $Ac = u$ and if we calculate c through this equation we get the approximation error of (3):

$$min \left\| y_j - y \right\|_{L^2(I)} = \left\| \hat{y}_j - y \right\|_{L^2(I)} = \sqrt{\left(\left\| y \right\|_{L^2(I)} \right)^2 - c^T A c} \quad (6)$$

The approximation error of the global $\hat{y}_j$ approximation on I is:

$$\left\| \tilde{y}_j - y \right\|_{L^2(I)} = \sqrt{\left(\left\| y \right\|_{L^2(I)} \right)^2 + \tilde{c}^T A \tilde{c} - 2\tilde{c}^T u} \quad \text{with } \tilde{c}_k = \left\langle y, \phi_{j,k} \right\rangle_{L^2(R)}$$

(6) we get through (which is general for every vector c right):

$$\left(\left\|y_j - y\right\|_{L^2(I)}\right)^2 = \left(\left\|y\right\|_{L^2(I)}\right)^2 - 2\left\langle y, y_j\right\rangle_{L^2(I)} + \left(\left\|y_j\right\|_{L^2(I)}\right)^2$$

with $\left\langle y, y_j\right\rangle_{L^2(I)} = \left\langle y, \sum\limits_{l=k_{min}}^{k_{max}} c_k \cdot \phi_{j,l}\right\rangle_{L^2(I)} = \sum\limits_{l=k_{min}}^{k_{max}} c_k \cdot \left\langle y, \phi_{j,l}\right\rangle_{L^2(I)} = c^T u$ and

$$\left(\left\|y_j\right\|_{L^2(I)}\right)^2 = \left\langle \sum\limits_{l=k_{min}}^{k_{max}} c_l \cdot \phi_{j,k}, \sum\limits_{k=k_{min}}^{k_{max}} c_k \cdot \phi_{j,k}\right\rangle_{L^2(I)} = \sum\limits_{l=k_{min}}^{k_{max}}\sum\limits_{k=k_{min}}^{k_{max}} c_l \cdot c_k \cdot \left\langle \phi_{j,l}, \phi_{j,k}\right\rangle_{L^2(I)} = c^T A c$$

So:

$$\left(\left\|y_j - y\right\|_{L^2(I)}\right)^2 = \left(\left\|y\right\|_{L^2(I)}\right)^2 - 2c^T u + c^T A c$$

For the approximation $\hat{y}_j$ (where we get c through $Ac = u$) we get equation (6).

With a special scalar product $\left\langle x, y\right\rangle_A := x^T A x$ and the induced norm (with positive definite A) $\left\|x\right\|_A = \sqrt{\left\langle x, y\right\rangle_A}$ we get:

$$\left(\left\|\tilde{y}_j - y\right\|_{L^2(I)}\right)^2 - \left(\left\|\hat{y}_j - y\right\|_{L^2(I)}\right)^2 = c^T A c + \tilde{c}^T A \tilde{c} - 2\tilde{c}^T u = c^T A c + \tilde{c}^T A \tilde{c} - 2\tilde{c}^T A c$$

$$= \left\langle c, c\right\rangle_A + \left\langle \tilde{c}, \tilde{c}\right\rangle_A - 2\left\langle c, \tilde{c}\right\rangle_A = \left\|\tilde{c} - c\right\|_A^2$$

For example:
The orthogonal projection of y on S_6 or V_6 is very near to y, because the differences of the $L^2(R)$ norm the $||Y|| - ||Y_j||$ is very small.

With the direct approximation through

(4) $$min\quad Q(c) = \sum\limits_{i=0}^{m} (y_j'(t_i) - y(t_i))^2$$

We set $k_{max} = 80$ and $j = 6$ ($r = 8$) and start the minimization:

```
fe[t_]:=  50/(1+50t²)
WCollocationS2Alg[y[t]-fe[t],-1,fe[-1]//N,-1.,1.,80,6,8,2]
```

Here the algorithm stops directly with $k_{max} = 80$, $j = 6$, $r = 8$ and a very small $Q_{min} = 6.98752\times10^{-22}$ and $Q_2 = 4.99871\times10^{-21}$. We started with bigger parameters to get less steps. The difference of the orthogonal projection $\tilde{y}_j$ from y on V_j and y (with $k_{max} = 80$) has the following graph:

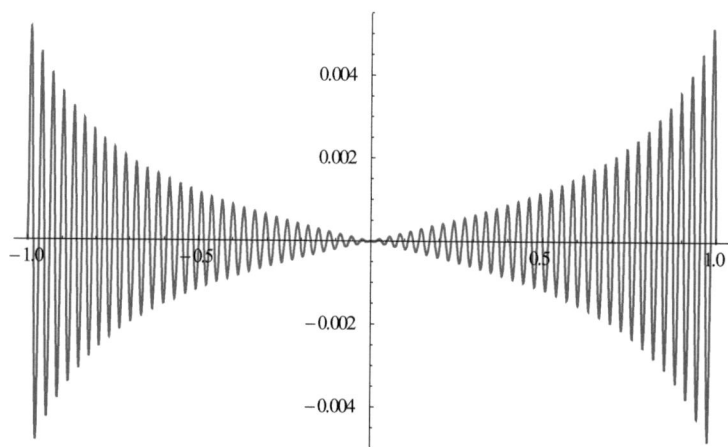

Here we get:

$$\sum_{i=0}^{m}(\tilde{y}_6(t_i)-y(t_i))^2 = 0.00131592 \quad \text{with} \quad h=\frac{t_{end}-t_0}{m} \quad \text{with} \quad m = r\cdot|k_{max}^{(0)}| = 80\cdot 8.$$

If we set $k_{max} = 1000$ we get with the same t_i the following sum of squares

$$\sum_{i=0}^{m}(\tilde{y}_6(t_i)-y(t_i))^2 = 7.42953\times 10^{-13}$$

which is small but even much bigger than the sum of squares with $k_{max} = 80$ and the direct approximation y_6:

$$\sum_{i=0}^{m}(y_6(t_i)-y(t_i))^2 = 6.98752\times 10^{-22}$$

Here are the graphs of $y_6 - y$ ($k_{max} = 80$):

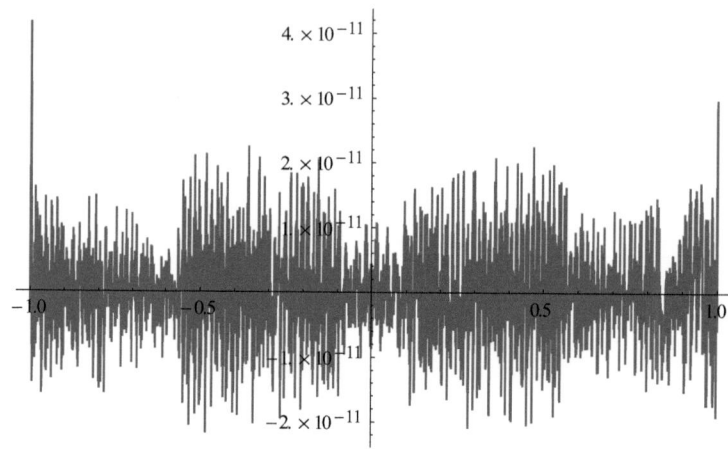

Finally here are some graphs of $y_j - y$ for selected combinations of j, k_{max} and r. Here we can see, that not only Q_{min} but Q_2 must be small, too,. That is what the algorithm does. With a too small r we get a big Q_2. In some cases Q_2 can be large because of big deviations at the edge of the approximation interval. Q_2 was theoretically studied in [16]. In that example we got in many simulations the first good approximations for a minimal value for j of 6 and for k_{max} the minimal value have been 70.

For $j = 6$, $k_{max} = 72$, and $r = 2$ we got a $Q_{min} = 1.82438\times 10^{-13}$ and $Q_2 = 9.04114\times 10^7$:

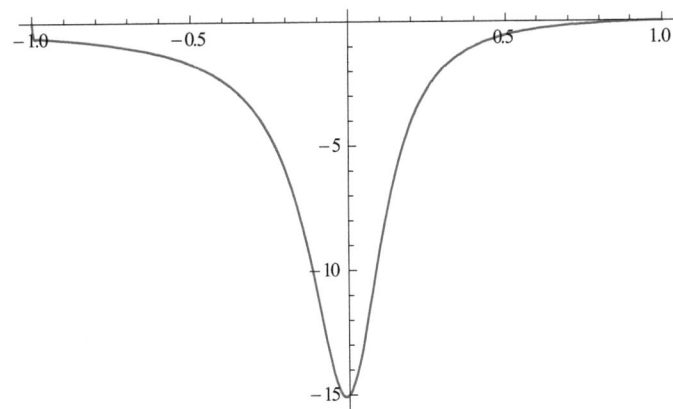

Here Q_2 was too big and so we got a bad approximation. The following examples have decreasing values of Q_2 and the approximation will get successively better.

For $j = 6$, $k_{max} = 72$, and $r = 3$ we got a $Q_{min} = 3.31586 \times 10^{-12}$ and $Q_2 = 5.98751 \times 10^{-6}$:

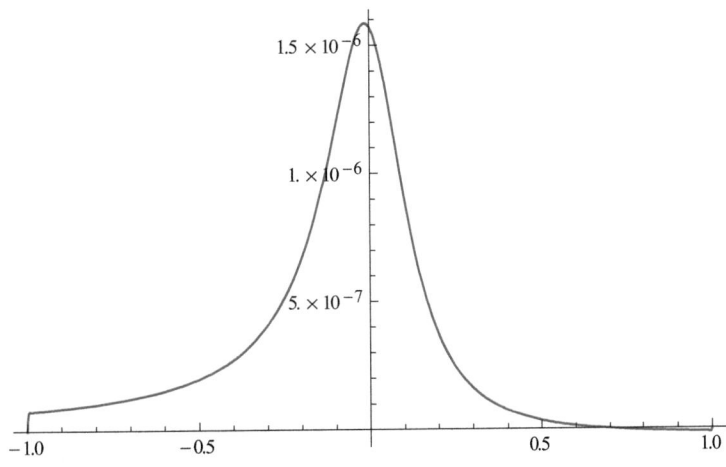

For $j = 6$, $k_{max} = 74$, and $r = 3$ we got a $Q_{min} = 3.35466 \times 10^{-12}$ and $Q_2 = 3.62539 \times 10^{-7}$:

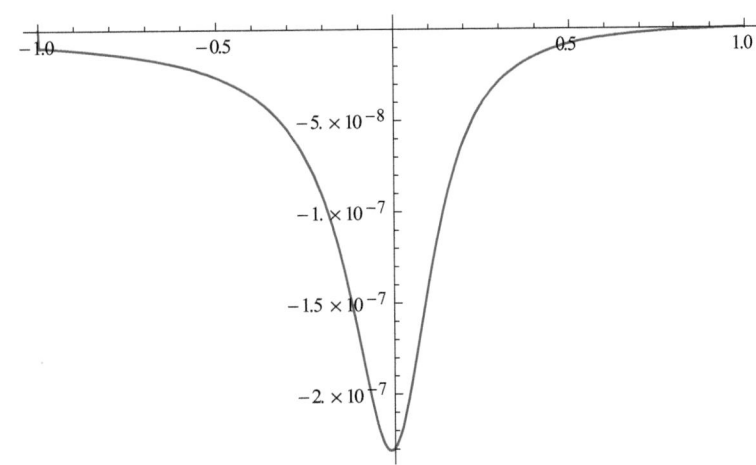

For $j = 6$, $k_{max} = 74$, and $r = 5$ we got a $Q_{min} = 4.74696 \times 10^{-12}$ and $Q_2 = 7.57142 \times 10^{-9}$:

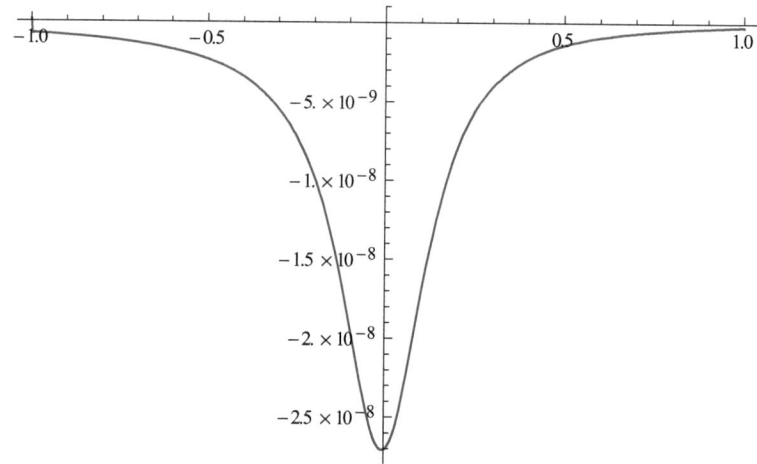

For $j = 6$, $k_{max} = 74$, and $r = 8$ we got a $Q_{min} = 7.71771 \times 10^{-12}$ and $Q_2 = 5.44366 \times 10^{-11}$:

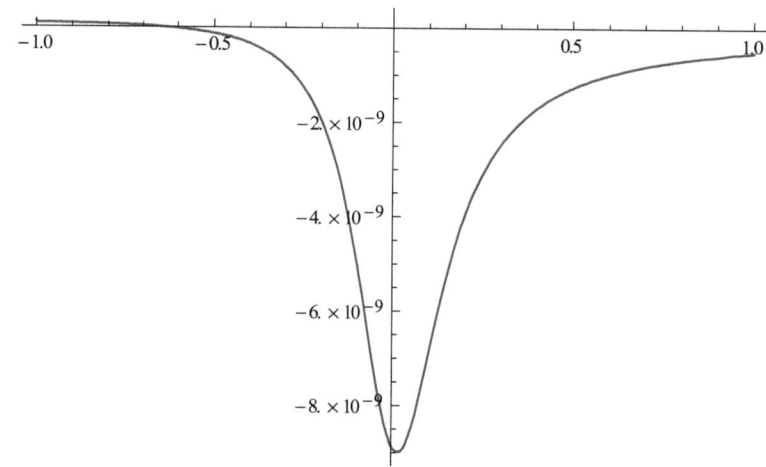

Here we can see how with decreasing values of Q_2 and with Q_{min} in the same magnitude the approximation error decreases.

Example V

We apply the algorithm on a second order ODE with boundary conditions:

$y'' = (y - (\mu \cdot \pi^2 + 1) \cdot cos(\pi t)) / \mu$ with $\mu = 1/100$ and with $y(-1) = -1$, $y(1) = -1$, the approximation interval is $I = [-1, 1]$.

We called the module for the algorithm with:

```
WCollocationS2Alg[-100(-(1+π²/100) Cos[π t]+y[t])+y″[t],{-1,1},{-1,-1},
-1.,1.,15,1,2,2]
```

Mathematica NDSolve has problems:
```
NDSolve::bvluc :
The equations derived from the boundary conditions are numerically ill-conditioned.
The boundary conditions may not be sufficient to uniquely define a solution. The
computed solution may match the boundary conditions poorly.
NDSolve::berr: There are significant errors _{0.,-1.76873×10⁻⁷}_ in the boundary
value residuals. Returning the best solution found. ≫
```

Here is the graph of $\eta - y$ (η is the NDSolve solution), were we can see, that the numerical solution of NDSolve has big deviations:

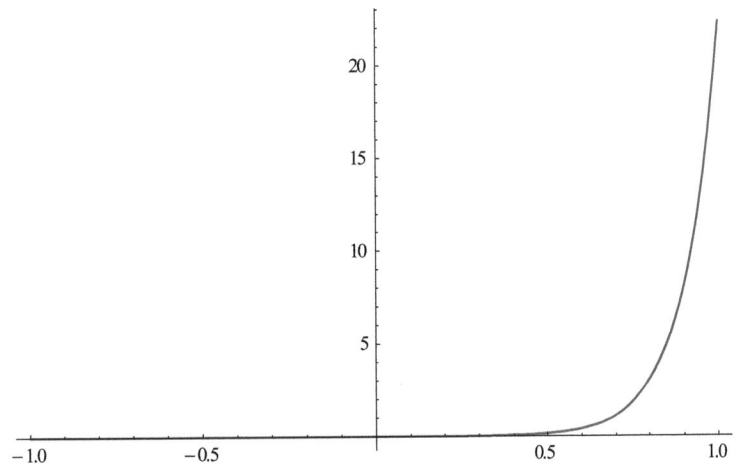

Here is the graph of η.

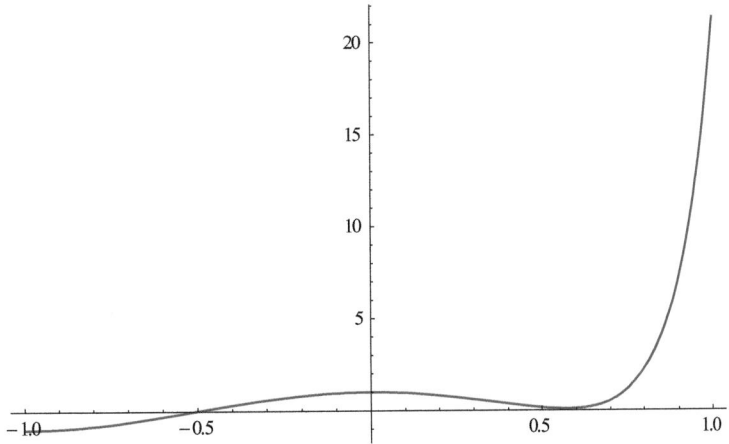

Now we see the iteration-protocol:

$k_{max}^{(0)}$	j	r	Q_{min}	Q_a
15	1	2	6.092×10^{-27}	1.08892×10^{-22}

Here are the graphs of y_1 and y (we see no differences):

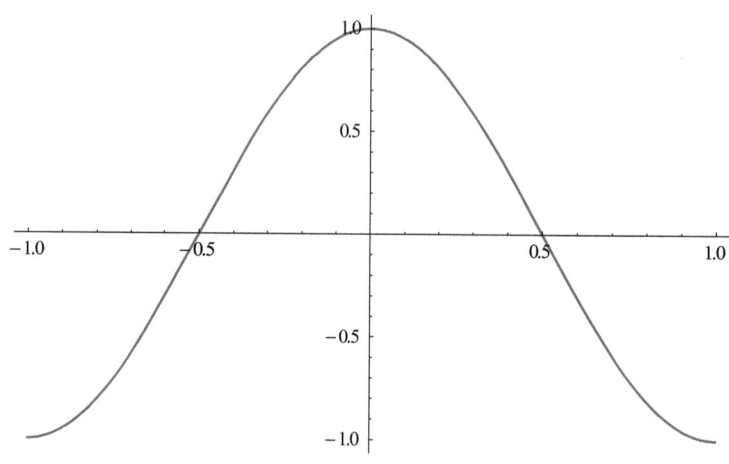

Here is the graph of $y_1 - y$:

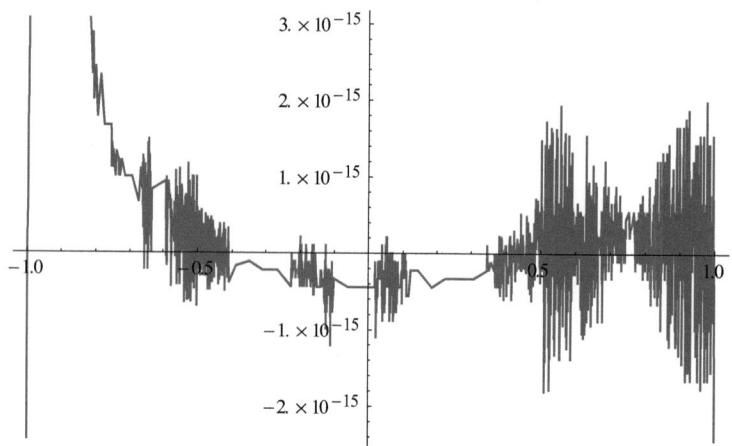

Now we see graphically the relation between a und Q_a/a in this example:

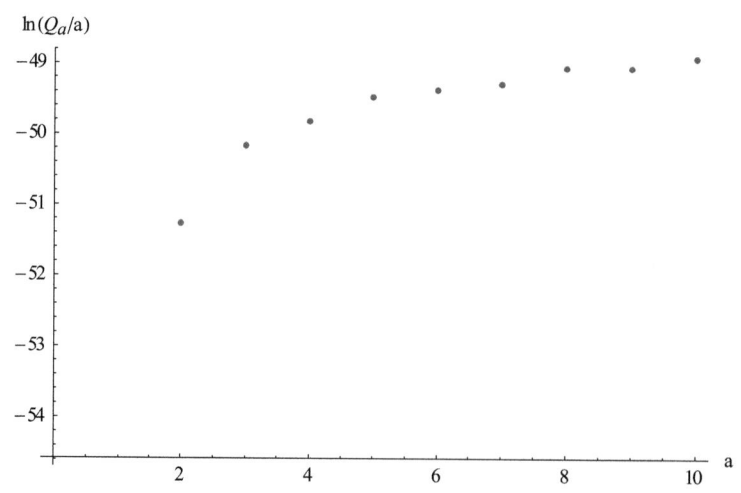

Even with $\mu = 1/1000$ we get after one step a very good approximation:

The algorithm stops directly with a small $k_{max} = 15$, $j = 1$, $r = 2$, $Q_{min} = 1.0731 \times 10^{-24}$ and $Q_a = 1.5481 \times 10^{-21}$.

Here is the graph of $y_1 - y$:

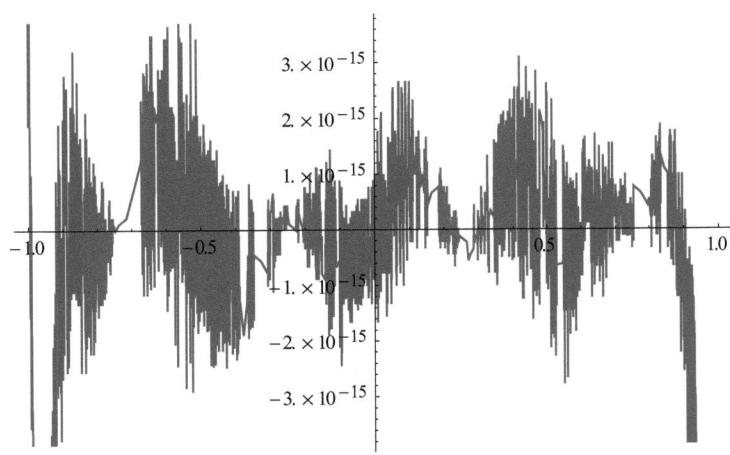

Here is the graph of $\eta - y$ (η is the NDSolve solution), were we can see, that the numerical solution of NDSolve has very big deviations:

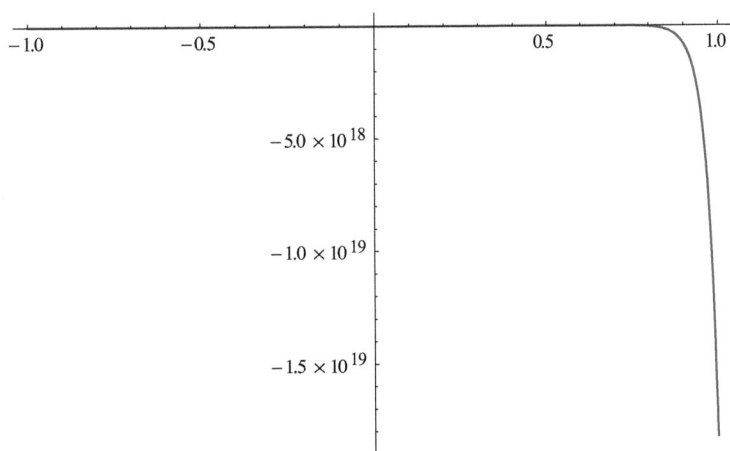

Example VI

We apply the algorithm on a second order ODE with boundary conditions:

$y'' = (y + y^2 - e^{-2t/sqrt(\mu)})/\mu$ with $\mu = 1/100$ and with $y(0) = 1$, $y(1) = e^{-1/sqrt(\mu)}$, the approximation interval is $I = [0, 1]$.

We called the module for the algorithm with:

```
WCollocationS2Alg[-100 (-e^-20t+y[t]+y[t]^2) + y''[t], {0,1}, {1,1/e^10},0.,1.,
15,1,2,2]
```

Mathematica NDSolve has problems:
```
NDSolve::ndsz: At _t_ == _0.9415179282009`_, step size is effectively zero;
singularity or stiff system suspected. ≫
General::stop: Further output of _NDSolve::ndsz_ will be suppressed during this
calculation. ≫ Divide::infy: Infinite expression _-(2.07326×10^-289/0.)_ encountered.
```

Mathematica automatically quits the kernel. The algorithm has no problems.

Now we see the iteration-protocol:

$k_{max}^{(0)}$	J	r	Q_{min}	Q_a
15	1	2	8.24021×10^{-14}	7.66528×10^{-11}

For critical examples we could start with a higher k_{max}, j and r.

Here are the graphs of y_4 and y (we see no differences):

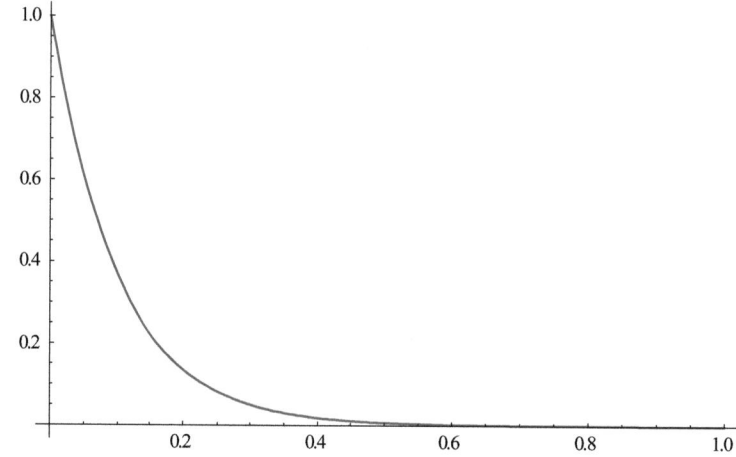

Here is the graph of $y_4 - y$:

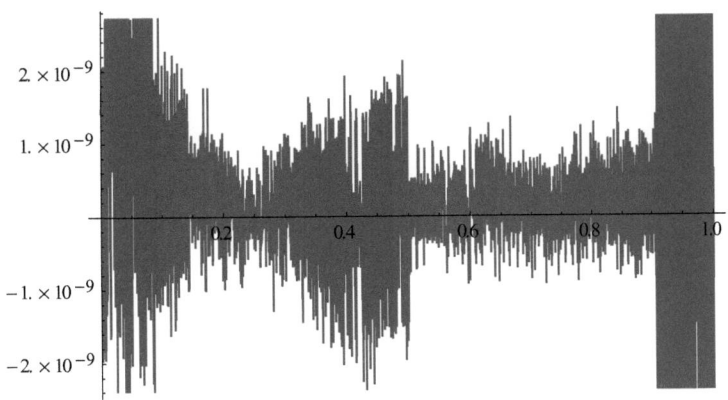

At last we see graphically the relation between a und Q_a/a in this example:

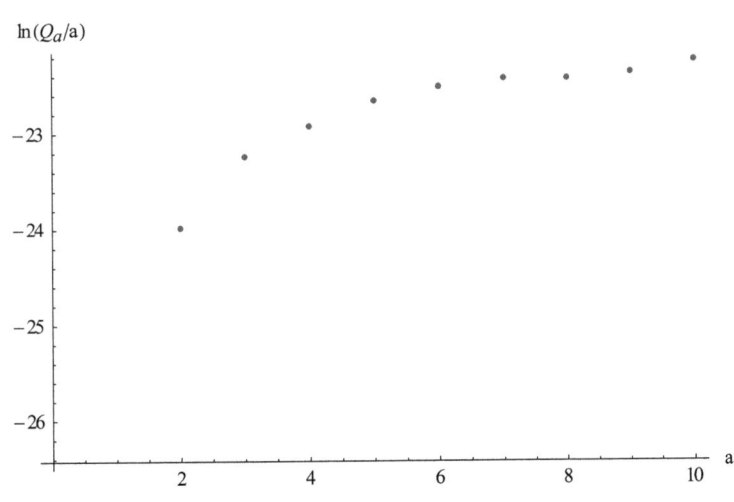

References

[1] Abdella, K. (2012). "Numerical Solution of Two-Point Boundary Value Problems Using Sinc Interpolation", *Proceedings of the American Conference on Applied Mathematics (American-Math '12): Applied Mathematics in Electrical and Computer Engineering*

[2] Ascher, U. A. Mattheij, R. M. M. Russell, R. D. (1988). „Numerical Solution of Boundary Value Problems for ODEs", *Prentice Hall (Series in Computational Mathematics)*

[3] Ascher, U. Christiansen, J. Russell, R. (1981). "Collocation Software for Boundary Value ODEs", *ACM Trans. Math. Software*

[4] Bertoluzza S. (2006). "Adaptive Wavelet Collocation Method for the Solution of Burgers Equation," *Transport Theory and Statistical Physics*

[5] Carlson, T. S. Dockery, J. Lund, J. (1997). "A Sinc-Collocation Method for Initial Value Problems", *Mathematics and Computation, Vol. 66, No. 217*

[6] Donoho, D. L.; (1992). "Interpolating Wavelet Transforms," *Tech. Rept. 408. Department of Statistics, Stanford University, Stanford*

[7] Hairer, E. Wanner, G. (1993). Vol. 1 : "Nonstiff Problems", *Springer 2. Auflage*

[8] Hairer, E. Wanner, G. (1996). Vol. 2 : "Stiff and Differential-Algebraic Problems", *Springer 2. Auflage*

[9] Mei, S.-L. Lv, H.-L. Ma, Q. (2008). „Construction of Interval Wavelet Based on Restricted Variational Principle and Its Application for Solving Differential Equations", *Hindawi Publishing Corporation Mathematical Problems in Engineering*

[10] Nurmuhammada, A. Muhammada, M., Moria, M. Sugiharab, M. (2005). "Double Exponential Transformation in the Sinc-Collocation Method for a Boundary Value Problem with Fourth-Order Ordinary Differential Equation," *Journal of Computational and Applied Mathematics*

[11] Qian, L. (2002). "On the Regularized Whittaker-Koltel'nikov-Shannon Sampling Theorem", *Proceedings of the Amarican Mathematical Society, Vol. 131, No. 4*

[12] Robertson, H. H. (1975). "Some Properties of Algorithms for Stiff Differential Equations", *J. Inst. Math. Applics.*

[13] Russell, R. D. Christiansen, J. (1979). "A Collocation Solver for Mixed Order Systems of Boundary Value Problems", *Mathematics of Computation*

[14] Schuchmann, M. (2012). "Approximation and Collocation with Wavelets. Approximations and Numerical Solving of ODEs, PDEs and IEs," *Osnabrück: DAV*

[15] Schuchmann, M. (2008). "Parameteridentifikation dynamischer Systeme auf günstigen Pfaden", *DAV*

[16] Schuchmann, M.; Rasguljajew, M. (2013). Error Estimation of an Approximation in a Wavelet Collocation Method. *Journal of Applied Computer Science & Mathematics, No. 14 (7) / 2013, Suceava*

[17] Schuchmann, M.; Rasguljajew, M. (2013). Parameter Identification with a Wavelet Collocation Method in a Partial Differential Equation. *Journal of Approximation Theory and Applied Mathematics (JATAM) Vol. 1*

[18] Schuchmann, M.; Rasguljajew, M. (2013). An Approach for a Parameter Estimation with a Wavelet Collocation Method. *Journal of Approximation Theory and Applied Mathematics (JATAM) Vol. 1*

[19] Schuchmann, M.; Rasguljajew, M. (2013). Approximation of Non $L^2(R)$ Functions on a Compact Interval with a Wavelet Base. *Journal of Approximation Theory and Applied Mathematics (JATAM) Vol. 2*

[20] Shi, Z.; Kouri, D.J.; Wei, G.W.; Hoffman, D. K.; (1999). „Generalized Symmetric Interpolating Wavelets", *Computer Physics Communications*

[21] Strang, G.; (1989). "Wavelets and Dilation Equations: A Brief Introduction", *SIAM Review Vol. 31, No. 4*

[22] Unser, M. (1996). "Vanishing Moments and the Approximation Power of Wavelet Expansions", *Proceedings of the 1996 IEEE International Conference on Image Processing*

Appendix

On www.jatam.de you will find the complete issues of volume 1 to volume 5. Here are the contents of the previous issues:

2013 Vol. 1
Contents:

An Approximation on a Compact Interval Calculated with a Wavelet Collocation Method can Lead to Much Better Results than other Methods
http://jatam.de/v5.pdf

Parameter Identification with a Wavelet Collocation Method in a Partial Differential Equation
http://jatam.de/v6.pdf

An Approach for a Parameter Estimation with a Wavelet Collocation Method
http://jatam.de/v8.pdf

Notes on Nonparametric Regression with Wavelets
http://jatam.de/v9.pdf

Extrapolation and Approximation with a Wavelet Collocation Method for ODEs
http://jatam.de/v10.pdf

2013 Vol. 2:
Contents:

Solving ODEs and DAEs with a Wavelet Collocation Method with Examples from the Chemical Reaction Kinetics
http://jatam.de/Art1-Vol-2-2013.pdf

Solving Integral Equations with a Wavelet Collocation Approach
http://jatam.de/Art2-Vol-2-2013.pdf

Approximation of Non $L^2(R)$ Functions on a Compact Interval with a Wavelet Base
http://jatam.de/Art3-Vol-2-2013.pdf

Comparing Approximations of a Wavelet Collocation Method of Various Wavelets
http://jatam.de/Art4-Vol-2-2013.pdf

2014 Vol. 3:
Contents:

Parameter Identification with a Wavelet Collocation Method for ODEs and DAEs
http://jatam.de/Art1-Vol-3-2014.pdf

Parameter Identification with a Wavelet Collocation Method in the Black Scholes Equation
http://jatam.de/Art2-Vol-3-2014.pdf

Adapted Linear Approximation for Logarithmic Kernel Integrals
http://jatam.de/Art3-Vol-3-2014.pdf

Identifying a Superposition with Trigonometric Functions by Applying a MRA with the Shannon Wavelet
http://jatam.de/Art4-Vol-3-2014.pdf

2014 Vol. 4:
Contents:

Approximation Error by Using a Finite Number of Base Coefficients for Special Types of Wavelets
http://jatam.de/Art1-Vol-4-2014.pdf

Solving Fredholm Integral Equations with Application of the Four Chebyshev Polynomials
http://jatam.de/Art2-Vol-4-2014.pdf

Here are more articles on the topic "approximation with wavelets" from the editors:

M. Schuchmann, M. Rasguljajew; (2013). *Error Estimation of an Approximation in a Wavelet Collocation Method.* Journal of Applied Computer Science & Mathematics, No. 14 (7) / 2013, Suceava. (http://jacs.usv.ro/index.php?pag=showcontent&issue=14&year=2013).

M. Schuchmann, M. Rasguljajew; (2013). *Error Estimation and Assessment of an Approximation in a Wavelet Collocation Method.* American Journal of Computational Mathematics, Vol.3, No.2, June 2013.

M. Schuchmann, M. Rasguljajew; (2013). *Determination of Optimal Parameters in a Wavelet Collocation Method.* International Journal of Emerging Technology and Advanced Engineering, Vol. 3, Issue 5, May 2013.
(http://www.ijetae.com/files/Volume3Issue5/IJETAE_0513_01.pdf)

M. Schuchmann, M. Rasguljajew; (2013). *Vergleich der Approximationsgüte verschiedener Wavelets zur numerischen Lösung gewöhnlicher Differentialgleichungen.*
Friedberger Hochschulschrift Band 33

M. Schuchmann, M. Rasguljajew; (2013). *Implementation and Testing an Algorithm for a Wavelet Collocation Method in Mathematica.* International Journal of Emerging Technology and Advanced Engineering, Vol. 3, Issue 6, June 2013. (http://www.ijetae.com/files/Volume3Issue6/IJETAE_0613_01.pdf)

M. Schuchmann, M. Rasguljajew; (2014). *Fourier Properties of Approximations of Functions on a Compact Interval with Daubechies Wavelets.* International Journal of Emerging Technology and Advanced Engineering, Vol. 4, Issue 4, April 2014. (http://www.ijetae.com/files/Volume4Issue4/IJETAE_0414_46.pdf)